Charles Hartwell Horne Cheyne

An elementary treatise on the planetary theory

With a collection of problems

Charles Hartwell Horne Cheyne

An elementary treatise on the planetary theory
With a collection of problems

ISBN/EAN: 9783337276577

Printed in Europe, USA, Canada, Australia, Japan

Cover: Foto ©berggeist007 / pixelio.de

More available books at **www.hansebooks.com**

AN

ELEMENTARY TREATISE

ON

THE PLANETARY THEORY,

WITH A COLLECTION OF PROBLEMS.

BY

C. H. H. CHEYNE, B.A.
SCHOLAR OF ST JOHN'S COLLEGE, CAMBRIDGE.

Cambridge:
MACMILLAN AND CO.
AND 23, HENRIETTA STREET, COVENT GARDEN,
London.

1862.

PREFACE.

IN this volume, an attempt has been made to produce a Treatise on the Planetary Theory, which, being elementary in character, should be so far complete, as to contain all that is usually required by students in this University. But it is not without diffidence that I submit my volume to their notice. In the earlier part of it, the methods which have been adopted are to some extent original*, and the general arrangement of the second chapter will, it is believed, be found to be new. Through the kindness of the Publishers, a portion of Pratt's *Mechanical Philosophy* has been placed at my disposal. Of this I have availed myself, particularly in the chapter on the Stability of the Planetary System; but, on the whole, comparatively little has been reprinted verbatim from that work. Among other sources of information, my obligations are mainly due to Pontécoulant's *Théorie Analytique du Système du Monde*, Airy's *Mathematical Tracts*, and Frost's *Planetary Theory* in the *Quarterly Journal of Mathematics:* but I have also referred to the

* Some of these have already appeared in Mathematical Journals.

Mécanique Céleste, the *Mécanique Analytique*, Mrs Somerville's *Mechanism of the Heavens*, (a work forming a complete Mathematical treatise on Physical Astronomy,) a Memoir by Prof. Donkin on the *Differential Equations of Dynamics*, *Phil. Trans.* 1855, &c. A collection of Problems has been added, taken chiefly from the Smith's Prize and Senate-House Examination Papers of the last twenty years. In conclusion, I would express my sincere thanks to Messrs A. Freeman, P. T. Main, and other friends, of St John's College, for the valuable assistance which they have afforded me, and would venture to hope that the work will be found useful.

C. H. H. CHEYNE.

St John's College,
October 9, 1862.

CONTENTS.

CHAPTER I.

INTRODUCTION.

ART.		PAGE
1.	Necessity of approximate methods	1
2.	Deviations of the planets from elliptic motion	,,
3.	Elements of the orbit	2
4.	Plane of the orbit	,,
5.	The Sun and planets considered to attract as if they were collected into their respective centres of gravity	,,
6.	Principle of Superposition of Small Motions, when admissible	3
7.	Difference between the Lunar and Planetary Theories	,,
8.	Component of the disturbing force in any direction.—Disturbing function	,,
9.	Meaning of the symbol $\dfrac{dR}{ds}$	5
10.	Disturbing function independent of any particular system of co-ordinates that may be employed	6
11.	Transformation of the expression for R'	7
12.	To explain how R may be expressed in terms of the time and the elements of the orbit	8
14.	Relations between the partial differential coefficients of R	10

CHAPTER II.

FORMULÆ FOR CALCULATING THE ELEMENTS OF THE ORBIT.

ART.		PAGE
18.	Equations of motion..	14
19.	Definition of the term *fixed in the plane of the orbit*	16
21.	Principle of the method of the Variation of Parameters ...	17
22.	Application of this method to the equations of motion......	18
23.	Definition of the *instantaneous ellipse*	19
24.	To obtain formulæ for calculating the elements.—Process explained, 24.—Mean distance, 25, 26.—Excentricity, 27, 28.—Longitude of perihelion, 29.—Node and inclination, 30—35.—Epoch, 36, 37.—Mean motion, 38	20
39.	Recapitulation of formulæ for calculating the elements.....	35

CHAPTER III.

DEVELOPMENT OF THE DISTURBING FUNCTION.

42.	Expansions of r_1, r_1', θ_1, θ_1', z, and z'	37
43.	Substitution of these in the expression for R	40
46.	Form of the terms in the development of R	43
47.	Determination of that part of R which is independent of the time explicitly	,,
50.	Order of magnitude of the periodical terms....................	45
52.	Proof that $(a^2 + a'^2 - 2aa' \cos \phi)^{-1}$ can be expanded in a series of cosines of multiples of ϕ	48
53.	Calculation of the coefficients	49
58.	Simplification of the expression for F	56

CHAPTER IV.

SECULAR VARIATIONS OF THE ELEMENTS OF THE ORBIT. STABILITY OF THE PLANETARY SYSTEM.

ART.		PAGE
60.	Definition of the term *secular*	60
61.	Formulæ for calculating the secular variations	"
62.	Approximate method of calculation	61
63.	Stability of the planetary system;—64, of the mean distances;—65—68, of the excentricities and inclinations	62

CHAPTER V.

SECULAR VARIATIONS OF THE ELEMENTS CONTINUED. INTEGRATION OF THE DIFFERENTIAL EQUATIONS.

72.	Integration of the equations for the excentricity and longitude of perihelion	69
74.	Stability of the excentricities in the case of two planets ...	72
76.	Examination of the expression for the longitude of perihelion	73
77.	When the apsidal line oscillates, to find the extent and periods of its oscillations	74
78.	Geometrical interpretation of the equations which give the secular variations of the excentricity and longitude of perihelion..	75
79.	Integration of the equations for the inclination and longitude of the node...	77
80.	Stability of the inclinations in the case of two planets	79
81.	Examination of the expression for the longitude of the node	80

ART.		PAGE
82.	When the line of nodes oscillates, to find the extent and periods of its oscillations	80
83.	Inclination of the orbits of two mutually disturbing planets to each other approximately constant......................	82
84.	Geometrical interpretation of the equations which give the secular variations of the node and inclination.........	,,
85.	Integration of the equation for the longitude of the epoch.	85
86.	Secular acceleration of the Moon's mean motion	86
87.	Formulæ for calculating the node and inclination of the plane of a planet's orbit relatively to the true ecliptic...	,,

CHAPTER VI.

PERIODIC VARIATIONS OF THE ELEMENTS OF THE ORBIT.

89.	Definition of the term *Periodical Variations*	89
91.	Expressions for the periodical variations of the elements	90
92.	Long inequalities...	91
93.	To select such terms in R as will produce the principal inequalities of long period..	91
94.	Relation between corresponding terms of the long inequalities in the mean motions of two mutually disturbing planets	92
95.	Variations of elements whose periods are very long	96
96.	Distinction between secular and periodic variations.........	97
98.	Periodic variations in radius vector	99
99.	Periodic variations in longitude	,,
100.	Example of the processes of this Chapter	100

CHAPTER VII.

DIRECT METHOD OF CALCULATING THE INEQUALITIES IN RADIUS VECTOR, LONGITUDE, AND LATITUDE.

ART.		PAGE
102.	Methods of Lagrange and Laplace	105
103.	Equations of motion	"
105.	Equation for the perturbation in radius vector	106
106.	Equation for the perturbation in longitude	108
107.	Equation for the perturbation in latitude	109
108.	Integration of the equation for the perturbation in radius vector	110
109.	First approximation to the value of δr	111
110.	Omission of the arbitrary term	113
112.	Certain terms to be neglected	114
113.	Second approximation to the value of δr	115
114.	Calculation of perturbations in longitude	116
115.	Determination of the constant g	117
116.	Long inequalities	118
117.	Integration of the equation for the perturbation in latitude	119

CHAPTER VIII.

ON THE EFFECTS WHICH A RESISTING MEDIUM WOULD PRODUCE IN THE MOTIONS OF THE PLANETS.

118.	Possibility of the existence of a very rare medium	121
119.	Equations of motion	122

CONTENTS.

ART.		PAGE
120.	Formula for calculating the mean distance......................	122
121.	Formula for calculating the excentricity	123
122.	Formula for calculating the longitude of perihelion	124
123.	Transformation of the above formulæ	125
124.	Formula for calculating the epoch	126
125.	Assumed form of the density	127
126.	Effect of the medium upon the elements, the orbit being supposed nearly circular	128
127.	The medium, though insensible to the planets, may yet affect the motions of comets...................................	129

PROBLEMS 130

APPENDIX.

ON THE FORM OF THE EQUATIONS OF ART. 39..................	140
ON THE GEOMETRICAL INTERPRETATION OF THE FORMULÆ FOR THE SECULAR VARIATIONS OF THE NODE AND INCLINATION	145
ON THE METHODS OF CALCULATING THE MASSES OF THE PLANETS	146

ERRATA.

Page 80, line 9, for *excentricities* read *inclinations*.

" " for $\dfrac{2\pi}{\pm h}$ read $\dfrac{2\pi}{\pm h_1}$.

THE PLANETARY THEORY.

CHAPTER I.

INTRODUCTION.

1. To determine the motion of a system of bodies subject only to their mutual attractions, is a problem the mathematical difficulties of which have not yet been overcome: hence in the particular cases of this problem which the Lunar and Planetary Theories present, recourse must be had to methods of approximation. Happily the arrangement of the Solar System renders approximate methods possible, and in the skilful hands of the Mathematicians of the last century, they have been brought to a high state of perfection.

2. If the Sun were the only attracting body, the planets would describe exact ellipses, agreeably to Kepler's first law; but in consequence of the attractions of the planets themselves, slight deviations from elliptic motion are produced. The method of calculating these deviations, to which our attention will chiefly be directed, is due to Euler; it consists in supposing the planets to move in ellipses, the elements (or arbitrary constants) of which are continually though slowly changing*.

---

* The legitimacy of this hypothesis will appear when we come to treat of the equations of motion. See Arts. 21 and 22.

3. Now the elements of an elliptic orbit are (i) the *mean distance*, or semi-axis major, (ii) the *excentricity*, (iii) the *longitude of perihelion*, i.e. of the point of the orbit nearest to the Sun, (iv) the *longitude of the epoch**, or mean longitude at the epoch from which the time is reckoned, (v) the *inclination* of the plane in which the orbit lies to some fixed plane of reference, (vi) the *longitude of the ascending node*. Of these (i) and (ii) determine the magnitude of the orbit, (iii) determines its position in its own plane, (v) and (vi) determine the position of this plane, and (iv) has reference to the position of the body itself in its orbit.

If the planets moved accurately in ellipses, these would be constants: we must however be prepared to consider them as variable quantities, which it will be the object of the problem to determine. They are termed the *elements of the orbit*.

4. But further, not only is it found that the true orbit of a planet is not an ellipse, but that it is not even a plane curve, although the departure of the planet from the plane in which it is at any instant moving is extremely slow. We define as the *plane of the orbit* the plane containing the radius vector and direction of motion of the planet at the instant under consideration.

5. We shall suppose the Sun and planets so distant from each other that they may be considered to attract as if they were condensed into their respective centres of gravity; a supposition which would be rigorously true if these bodies were exactly spherical, and either of uniform density or composed of concentric spherical shells, the density of each shell being uniform throughout. The errors, however, thus introduced into the motions of translation are found to be inappreciable for the planets, though not in the case of their satellites. The motions of rotation will not be considered in the present treatise.

* Also briefly termed the *epoch*.

6. Moreover, since the masses of the planets are extremely small in comparison of that of the Sun, it follows that in cases where it is not necessary to carry the approximation beyond the first order of these masses, we are permitted to avail ourselves of the Principle of the Superposition of Small Motions, and thus to reduce the problem to a case of that of the Three Bodies.

7. So far the Theory of the Planets resembles that of the Moon, and the same method of treatment might be employed in both cases. But they differ in this respect: the ratio of the distances of the disturbed and disturbing bodies from the central one* is much smaller in the Lunar than in the Planetary Theory, so that if in the latter theory the approximation were made by means of series proceeding by powers of this ratio, it would be necessary to retain many more terms than are required in the former. For this reason a different method of development is employed. The perturbations of the Moon, however, are far larger than those of the planets, since in the former case the Sun, of which the mass is enormous, and the distance not proportionately great, is one of the disturbing bodies.

8. *To find an expression for the component in any direction of the force which disturbs the motion of a given planet relatively to the Sun.*

Let M denote the mass of the Sun, m, m', m'', &c., those of the planets, and suppose the relative motion of m required.

Let x, y, z, x', y', z', x'', y'', z'', &c., be the co-ordinates of m, m', m'', &c., referred to any system of rectangular axes

---

* By the *central body* is meant that whose attraction exercises the greatest influence on the body whose motion is required; the Sun for instance in the Theory of the Planets, and the Earth in that of the Moon. All the other attracting bodies are called *disturbing bodies.*

originating in the centre of gravity of the Sun; r, r', r'', &c., their distances from the origin; ρ', ρ'', &c., the distances of m', m'', &c., from m.

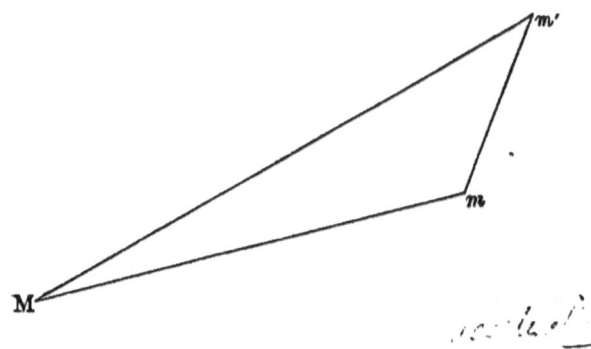

Now if to every body of the system we apply forces equal and opposite to those which act upon the Sun, we shall reduce the latter to rest without affecting the relative motion. Hence, considering the action of only one disturbing planet m', the forces acting upon m will be

$$\frac{M+m}{r^2} \text{ in direction } mM,$$

$$\frac{m'}{\rho'^2} \text{ in direction } mm',$$

$$\frac{m'}{r'^2} \text{ in direction } m'M,$$

of which the last two constitute the disturbing force.

Let $V = \dfrac{m'}{\rho'}$: then on the hypothesis of Art. 5, V will be the potential of m', and the components parallel to the axes of the disturbing force due to the action of m', will be

$$\frac{dV}{dx} - \frac{m'x'}{r'^3},$$

$$\frac{dV}{dy} - \frac{m'y'}{r'^3},$$

$$\frac{dV}{dz} - \frac{m'z'}{r'^3}.$$

Let s denote the length of the arc of any curve measured from some fixed point up to m*: then the resolved part of the disturbing force parallel to the tangent at m to this curve will be

$$\frac{dV}{ds} - \left(\frac{m'x'}{r'^3}\frac{dx}{ds} + \frac{m'y'}{r'^3}\frac{dy}{ds} + \frac{m'z'}{r'^3}\frac{dz}{ds}\right),$$

or restoring to V its value

$$\frac{d}{ds}\left\{\frac{m'}{\rho'} - \frac{m'}{r'^3}(xx' + yy' + zz')\right\},$$

which may be written $\frac{dR'}{ds}$, if

$$R' = \frac{m'}{\rho'} - \frac{m'}{r'^3}(xx' + yy' + zz').$$

If we express in like manner the disturbing forces due to the actions of m'', m''', &c., we shall have for the whole component in this direction

$$\frac{dR'}{ds} + \frac{dR''}{ds} + \ldots,$$

or $\frac{dR}{ds}$, if $R = R' + R'' + \ldots$.

The function R is called the *disturbing function*.

9. From the manner in which $\frac{dR}{ds}$ has been introduced, it appears that R is supposed to be expressed in terms of s and quantities which do not vary with s. It must however

* This arbitrary curve we shall term the *curve of reference*.

be borne in mind that in $\dfrac{dR}{ds}$ the variation is purely hypothetical, and has nothing whatever to do with the actual variation of R due to the motion of the planet.

For example, suppose the curve of reference a straight line parallel to the axis of x, and let R be expressed in terms of x, y and z; then in this case x only will vary, and the disturbing force parallel to the axis of x will be denoted by $\dfrac{dR}{dx}$, y and z being considered constant in the differentiation. Similarly, the disturbing forces parallel to the axes of y and z will be expressed by $\dfrac{dR}{dy}$ and $\dfrac{dR}{dz}$ respectively, the differential coefficients being strictly partial.

Again, suppose the curve of reference a circle with its plane parallel to that of xy, and its centre in the axis of z, and let R be expressed in terms of the polar co-ordinates (r_1, θ_1) of the projection of the planet on the plane of xy, and its distance (z) from this plane; then in this case θ_1 only will vary, and the disturbing force perpendicular to the projected radius vector will be expressed by $\dfrac{dR}{r_1 d\theta_1}$, r_1 and z being considered constant in the differentiation. Similarly, the forces parallel to the projected radius vector and to the axis of z, will be expressed by the partial differential coefficients $\dfrac{dR}{dr_1}$, $\dfrac{dR}{dz}$ respectively.

10. The disturbing function, like the potential, is independent of any particular system of co-ordinates that may be employed. For

$$R' = \frac{m'}{\rho'} - \frac{m'}{r'^3}(xx' + yy' + zz')$$

$$= \frac{m'}{\rho'} - \frac{m'r}{r'^2}\left(\frac{x}{r}\frac{x'}{r'} + \frac{y}{r}\frac{y'}{r'} + \frac{z}{r}\frac{z'}{r'}\right)$$

$$= \frac{m'}{\rho'} - \frac{m'r}{r'^2}\cos\omega,$$

if ω denote the inclination of r' to r.

11. *To express R' in terms of the polar co-ordinates of the projections of* m *and* m' *on a fixed plane, and of their distances from it.*

Take the fixed plane for that of xy: let r_1, r_1' be the projections of r, r' upon it, and θ_1, θ_1' the inclinations of r_1, r_1' to the axis of x; then

$$x = r_1\cos\theta_1, \quad y = r_1\sin\theta_1,$$
$$x' = r_1'\cos\theta_1', \quad y' = r_1'\sin\theta_1';$$

therefore
$$xx' + yy' + zz' = r_1 r_1'\cos(\theta_1 - \theta_1') + zz',$$
$$\rho'^2 = (x-x')^2 + (y-y')^2 + (z-z')^2$$
$$= r_1^2 + r_1'^2 - 2r_1 r_1'\cos(\theta_1 - \theta_1') + (z-z')^2,$$
$$r'^2 = x'^2 + y'^2 + z'^2$$
$$= r_1'^2 + z'^2.$$

Hence by substitution,

$$R' = \frac{m'}{\{r_1^2 + r_1'^2 - 2r_1 r_1'\cos(\theta_1 - \theta_1') + (z-z')^2\}^{\frac{1}{2}}}$$
$$- \frac{m'\{r_1 r_1'\cos(\theta_1 - \theta_1') + zz'\}}{\{r_1'^2 + z'^2\}^{\frac{3}{2}}}.$$

12. In a subsequent chapter we shall consider the development of R in terms of the time and the elements of the orbit, in a series ascending by powers and products of the excentricities and inclinations, which in the Planetary Theory

are very small. At present we shall content ourselves with shewing *how R* may be expressed in terms of these quantities. We shall assume that the equations connecting the co-ordinates, the time, and the elements in an elliptic orbit, hold also in the case of a disturbed planet.

13. *To explain how* R *may be expressed in terms of the time and the elements of the orbit.*

Let r, θ denote the radius vector and longitude of the disturbed planet, the latter being measured on a fixed plane of reference as far as the node, and thence on the plane of the orbit: let the elements be a the mean distance, e the excentricity, ϖ the longitude of perihelion, ϵ the longitude of the epoch, (the last two being measured in the same way as θ,) Ω the longitude of the node measured on the plane of reference, and i the inclination of the plane of the orbit to the plane of reference. Our object is to express R in terms of t and these elements.

Again, let θ_0, ϖ_0, ϵ_0, Ω_0 denote the longitudes of the planet, of perihelion, of the epoch, and of the node, measured entirely on the plane of the orbit.

Let a sphere be described with its centre coinciding with that of the Sun, and its radius of any magnitude: let the planes of reference and of the orbit cut it in the great circles NM, NP, then the line of nodes will cut it in N; let the radius vector of the planet cut it in P, the projection of this radius on the plane of reference in M, and the lines from which θ, θ_0 are measured in L, O respectively. We shall suppose L to be the same origin as that from which θ_1 is measured in Art. 11.

Then in the figure $LM = \theta_1$, $LN + NP = \theta$, $OP = \theta_0$, $LN = \Omega$, the angle $PNM = i$, and $PM =$ the latitude of the planet which we shall denote by λ.

INTRODUCTION.

Hence from the right-angled triangle PNM,

$$\tan(\theta_1 - \Omega) = \cos i \tan(\theta - \Omega) \quad \text{(1)},$$
$$\sin \lambda = \sin i \sin(\theta - \Omega) \quad \text{(2)}:$$

also
$$r_1 = r \cos \lambda \quad \text{(3)},$$
$$z = r \sin \lambda \quad \text{(4)}.$$

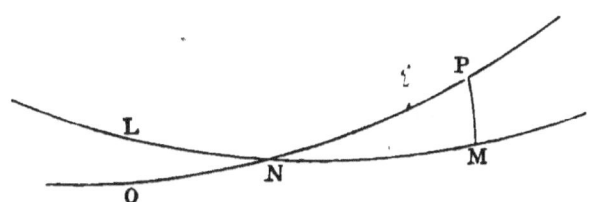

Again, from the formulæ of elliptic motion*,

$$r = a\{1 + \tfrac{1}{2}e^2 - e\cos(nt + \epsilon_0 - \varpi_0) - \tfrac{1}{2}e^2 \cos 2(nt + \epsilon_0 - \varpi_0) - \ldots\},$$

$$\theta_0 = nt + \epsilon_0 + 2e \sin(nt + \epsilon_0 - \varpi_0) + \tfrac{5}{4}e^2 \sin 2(nt + \epsilon_0 - \varpi_0) + \ldots \dagger:$$

but $\quad \theta - \theta_0 = LN - ON = \epsilon - \epsilon_0 = \varpi - \varpi_0;$

therefore $\quad \theta = \theta_0 + \epsilon - \epsilon_0, \quad \epsilon_0 - \varpi_0 = \epsilon - \varpi,$

and our formulæ become

$$r = a\{1 + \tfrac{1}{2}e^2 - e\cos(nt + \epsilon - \varpi) - \tfrac{1}{2}e^2 \cos 2(nt + \epsilon - \varpi) - \ldots\} \ldots (5),$$

$$\theta = nt + \epsilon + 2e \sin(nt + \epsilon - \varpi) + \tfrac{5}{4}e^2 \sin 2(nt + \epsilon - \varpi) + \ldots \ldots (6).$$

In Art. 11 we have expressed R' in terms of r_1, θ_1 and z; hence by equations (1) to (4) it may be expressed as a function of r, θ, Ω, and i: we may then substitute for r and θ from equations (5) and (6), and R' will be expressed in terms

---

* See *Tait and Steele's Dynamics*, Arts. 114 and 122.

† n is termed the mean motion, and is connected with the mean distance by the equation $n^2 a^3 = \mu$, where $\mu = M + m$.

of t and the elements of the orbit. Similarly R'', R''', &c., and therefore R may be expressed in terms of t and the elements.

14. We proceed to investigate certain relations which subsist between the partial differential coefficients of R with respect to the co-ordinates of the disturbed planet, and its partial differential coefficients with respect to the elements of the orbit. These will be useful in obtaining the formulæ by which the values of the elements are calculated.

We premise that when we speak of the partial differential coefficient of R with respect to one of the elements, we suppose R expressed in the manner indicated in the last article, and that the time as well as the other elements are considered constant in the differentiation: when we speak of the partial differential coefficient of R with respect to r or θ, we suppose R expressed in terms of r, θ, i and Ω, which may be done by equations (1) to (4).

15. *To shew that* $\dfrac{dR}{d\theta} = \dfrac{dR}{d\epsilon} + \dfrac{dR}{d\varpi}$.

Equations (5) and (6) of Art. 13 may be written

$$r = f(nt + \epsilon - \varpi),$$

$$\theta - \varpi = \phi(nt + \epsilon - \varpi),$$

whence it follows that

$$\frac{dr}{d\epsilon} + \frac{dr}{d\varpi} = 0,$$

$$\frac{d\theta}{d\epsilon} + \frac{d\theta}{d\varpi} = 1.$$

Now since ϵ and ϖ enter into R only through r and θ,

$$\frac{dR}{d\epsilon} = \frac{dR}{dr}\frac{dr}{d\epsilon} + \frac{dR}{d\theta}\frac{d\theta}{d\epsilon},$$

INTRODUCTION. 11

$$\frac{dR}{d\varpi} = \frac{dR}{dr}\frac{dr}{d\varpi} + \frac{dR}{d\theta}\frac{d\theta}{d\varpi};$$

therefore, by addition,

$$\frac{dR}{d\epsilon} + \frac{dR}{d\varpi} = \frac{dR}{d\theta}.$$

16. *To shew that* $\dfrac{dR}{d\theta_1} = \dfrac{dR}{d\epsilon} + \dfrac{dR}{d\varpi} + \dfrac{dR}{d\Omega}.$

From equations (1) to (4) of Art. 13 we obtain

$$r_1 = \phi(r, \theta - \Omega, i),$$
$$\theta_1 - \Omega = \chi(\theta - \Omega, i),$$
$$z = \psi(r, \theta - \Omega, i),$$

where ϕ, χ, ψ are symbols of functionality.

It follows that

$$\frac{dr_1}{d\theta} + \frac{dr_1}{d\Omega} = 0,$$

$$\frac{d\theta_1}{d\theta} + \frac{d\theta_1}{d\Omega} = 1,$$

$$\frac{dz}{d\theta} + \frac{dz}{d\Omega} = 0.$$

Now since by Art. 11, R is a function of r_1, θ_1, and z,

$$\frac{dR}{d\theta} = \frac{dR}{dr_1}\frac{dr_1}{d\theta} + \frac{dR}{d\theta_1}\frac{d\theta_1}{d\theta} + \frac{dR}{dz}\frac{dz}{d\theta},$$

$$\frac{dR}{d\Omega} = \frac{dR}{dr_1}\frac{dr_1}{d\Omega} + \frac{dR}{d\theta_1}\frac{d\theta_1}{d\Omega} + \frac{dR}{dz}\frac{dz}{d\Omega};$$

therefore, by addition,

$$\frac{dR}{d\theta} + \frac{dR}{d\Omega} = \frac{dR}{d\theta_1},$$

whence, by the last article,
$$\frac{dR}{d\theta_1} = \frac{dR}{d\epsilon} + \frac{dR}{d\varpi} + \frac{dR}{d\Omega}.$$

17. *To obtain* $\frac{dR}{de}$ *in terms of* $\frac{dR}{dr}$ *and* $\frac{dR}{d\theta}$.

If u denote the excentric anomaly, we have*

$$r = a(1 - e\cos u) \quad \dots \dots \dots \dots (1),$$

$$\tan\frac{\theta - \varpi}{2} = \sqrt{\left(\frac{1+e}{1-e}\right)} \tan\frac{u}{2} \quad \dots \dots \dots (2),$$

$$nt + \epsilon - \varpi = u - e\sin u \quad \dots \dots \dots \dots (3),$$

from which r and θ may be expressed in terms of t and the elements by eliminating u. Assuming r and θ so expressed, we proceed to obtain $\frac{dr}{de}$ and $\frac{d\theta}{de}$.

From (1), $\frac{dr}{de} = a\left(e\sin u\,\frac{du}{de} - \cos u\right),$

and from (3), $\frac{du}{de}(1 - e\cos u) - \sin u = 0 \quad \dots \dots \dots (4);$

eliminating $\frac{du}{de}$, we have

$$\frac{dr}{de} = a\left\{\frac{e\sin^2 u}{1 - e\cos u} - \cos u\right\},$$

$$= a\left\{\frac{e - \cos u}{1 - e\cos u}\right\},$$

$$= a\left\{\frac{1 - e\cos u - (1 - e^2)}{e(1 - e\cos u)}\right\},$$

* See *Tait and Steele's Dynamics*, Arts. 110, 111, and 122.

$$= \frac{a}{e}\left\{1 - \frac{a(1-e^2)}{r}\right\}, \text{ by (1)},$$

$$= -a\cos(\theta - \varpi),$$

by the polar equation to the ellipse.

Again, differentiating the logarithms of equation (2),

$$\frac{1}{\sin(\theta - \varpi)}\frac{d\theta}{de} = \frac{1}{2}\left(\frac{1}{1+e} + \frac{1}{1-e}\right) + \frac{1}{\sin u}\frac{du}{de};$$

eliminating $\dfrac{du}{de}$ by means of (4),

$$\frac{1}{\sin(\theta - \varpi)}\frac{d\theta}{de} = \frac{1}{1-e^2} + \frac{1}{1 - e\cos u},$$

$$= a\left\{\frac{1}{a(1-e^2)} + \frac{1}{r}\right\}, \text{ by (1)},$$

$$= a\left\{\frac{\mu}{h^2} + \frac{1}{r}\right\},$$

if $h^2 = \mu a(1-e^2)$; therefore

$$\frac{d\theta}{de} = a\left(\frac{\mu}{h^2} + \frac{1}{r}\right)\sin(\theta - \varpi).$$

Now since R is a function of e only because it is a function of r and θ,

$$\frac{dR}{de} = \frac{dR}{dr}\frac{dr}{de} + \frac{dR}{d\theta}\frac{d\theta}{de},$$

$$= -a\cos(\theta - \varpi)\frac{dR}{dr} + a\left(\frac{\mu}{h^2} + \frac{1}{r}\right)\sin(\theta - \varpi)\frac{dR}{d\theta}.$$

Since $\theta - \varpi = \theta_0 - \varpi_0$, (see Art. 13,) this equation may be written

$$\frac{dR}{de} = -a\cos(\theta_0 - \varpi_0)\frac{dR}{dr} + a\left(\frac{\mu}{h^2} + \frac{1}{r}\right)\sin(\theta_0 - \varpi_0)\frac{dR}{d\theta},$$

under which form it will be useful in the next chapter.

CHAPTER II.

FORMULÆ FOR CALCULATING THE ELEMENTS OF THE ORBIT.

18. We now proceed to form equations of motion, taking the Sun's centre for the origin of co-ordinates, the radius vector of the planet for the axis of x, a perpendicular to it in the plane of the orbit for the axis of y, and a normal to this plane for the axis of z. With this system, it will be shewn that two of the resulting equations can be expressed in the same forms as if the planet moved in one plane.

Let x, y, z be the co-ordinates, u, v, w the velocities of the planet with reference to three rectangular axes originating in the Sun's centre, and moving with angular velocities ϕ_1, ϕ_2, ϕ_3 about their instantaneous positions: let X, Y, Z be the accelerations due to the impressed forces in the directions of the axes. Then (Routh's *Rigid Dynamics*, Art. 114),

$$\left. \begin{aligned} u &= \frac{dx}{dt} - y\phi_3 + z\phi_2 \\ v &= \frac{dy}{dt} - z\phi_1 + x\phi_3 \\ w &= \frac{dz}{dt} - x\phi_2 + y\phi_1 \end{aligned} \right\} \dots\dots\dots\dots\dots (1),$$

and the equations of motion are

FORMULÆ FOR CALCULATING THE ELEMENTS.

$$X = \frac{du}{dt} - v\phi_3 + w\phi_2$$
$$Y = \frac{dv}{dt} - w\phi_1 + u\phi_3 \quad \quad \quad \ldots\ldots\ldots\ldots\ldots (2).$$
$$Z = \frac{dw}{dt} - u\phi_2 + v\phi_1$$

In these equations ϕ_1, ϕ_2, ϕ_3 are arbitrary; we propose so to determine them that the axis of x may coincide with the radius vector of the planet, and the plane of xy with the plane of the orbit.

In order that the axis of x may coincide with the radius vector of the planet, we must have

$$x = r, \quad y = 0, \quad z = 0,$$

always; and therefore

$$\frac{dx}{dt} = \frac{dr}{dt}, \quad \frac{dy}{dt} = 0, \quad \frac{dz}{dt} = 0:$$

and in order that the plane of xy may coincide with the plane of the orbit, we must have

$$w = 0$$

always; and therefore

$$\frac{dw}{dt} = 0.$$

Hence equations (1) give

$$u = \frac{dr}{dt}, \quad v = r\phi_3, \quad \phi_2 = 0;$$

and equations (2) become

$$X = \frac{d^2r}{dt^2} - r\phi_3^2,$$
$$Y = r\frac{d\phi_3}{dt} + 2\phi_3\frac{dr}{dt},$$
$$Z = r\phi_3\phi_1.$$

In order to reduce the first two of these equations to the forms they would take if the motion were in one plane, let $\phi_3 = \dfrac{d\theta_0}{dt}$: thus

$$X = \frac{d^2 r}{dt^2} - r\left(\frac{d\theta_0}{dt}\right)^2,$$

$$Y = \frac{1}{r}\frac{d}{dt}\left(r^2 \frac{d\theta_0}{dt}\right),$$

$$Z = r\phi_1 \frac{d\theta_0}{dt}.$$

19. If we measure on the plane of the orbit from the planet's radius vector in a direction contrary to that of motion an angle equal to θ_0, we arrive at what may be considered as the origin from which θ_0 is measured. Since this will be a point having no angular velocity about the axis of z, which is normal to the plane of the orbit, it is said to be *fixed in the plane of the orbit**.

20. We shall for the present confine our attention to the first two of the above equations.

In order to find the components of the disturbing force parallel and perpendicular to the radius vector of the planet, let us take first the radius vector as the curve of reference; then $s = r$, and R being supposed expressed as a function of r, θ, Ω, and i (see Art. 13), we have

$$\frac{dR}{ds} = \frac{dR}{dr}\frac{dr}{ds} + \frac{dR}{d\theta}\frac{d\theta}{ds} + \frac{dR}{d\Omega}\frac{d\Omega}{ds} + \frac{dR}{di}\frac{di}{ds}$$

$$= \frac{dR}{dr},$$

---

* It is necessary to define this term, since the definition of Art. 4 is not sufficient completely to regulate the motion of the plane of the orbit, though affording it a distinct geometrical position.

since θ, Ω and i do not vary with s. Hence the disturbing force in direction of the radius vector $= \dfrac{dR}{dr}$.

Again, let us take as the curve of reference a circle in the plane of the orbit, with its centre coinciding with that of the sun; then $\delta s = r\delta\theta$, and we have

$$\frac{dR}{ds} = \frac{1}{r}\frac{dR}{d\theta},$$

since r, Ω, and i do not vary with s. Hence the disturbing force perpendicular to the radius vector $= \dfrac{1}{r}\dfrac{dR}{d\theta}$.

We have then

$$X = -\frac{\mu}{r^2} + \frac{dR}{dr}, \quad Y = \frac{1}{r}\frac{dR}{d\theta},$$

and the equations become

$$\frac{d^2r}{dt^2} - r\left(\frac{d\theta_0}{dt}\right)^2 = -\frac{\mu}{r^2} + \frac{dR}{dr} \quad \dots\dots\dots\dots (1),$$

$$\frac{d}{dt}\left(r^2 \frac{d\theta_0}{dt}\right) = \frac{dR}{d\theta} \quad \dots\dots\dots\dots\dots (2).$$

21. These equations do not admit of rigorous integration, but we may reduce them by the method of the Variation of Parameters to a system of differential equations of the first order. The principle of this method may be explained as follows. Suppose it required to integrate the equations

$$\left.\begin{array}{l}\phi_1\left(x,\ y,\ t,\ \dfrac{dx}{dt},\ \dfrac{dy}{dt},\ \dfrac{d^2x}{dt^2},\ \dfrac{d^2y}{dt^2}\right) = P_1 \\[2mm] \phi_2\left(x,\ y,\ t,\ \dfrac{dx}{dt},\ \dfrac{dy}{dt},\ \dfrac{d^2x}{dt^2},\ \dfrac{d^2y}{dt^2}\right) = P_2\end{array}\right\} \dots (i),$$

where P_1, P_2 are functions of t. The solution of these equations can be made to depend upon that of the equations

C. P. T.

$\phi_1 = 0$, $\phi_2 = 0$. Suppose the four first integrals of these last equations to be

$$\left. \begin{array}{l} \chi_1\left(x,\ y,\ t,\ \dfrac{dx}{dt},\ \dfrac{dy}{dt}\right) = c_1 \\[4pt] \chi_2\left(x,\ y,\ t,\ \dfrac{dx}{dt},\ \dfrac{dy}{dt}\right) = c_2 \\[4pt] \chi_3\left(x,\ y,\ t,\ \dfrac{dx}{dt},\ \dfrac{dy}{dt}\right) = c_3 \\[4pt] \chi_4\left(x,\ y,\ t,\ \dfrac{dx}{dt},\ \dfrac{dy}{dt}\right) = c_4 \end{array} \right\} \ldots\ldots\ldots\ldots (ii),$$

where c_1, c_2, c_3, c_4 are arbitrary constants or parameters. The method of the *Variation of Parameters* consists in so determining c_1, c_2, c_3 and c_4 as functions of t, that these integrals (and therefore the two final integrals of the equations $\phi_1 = 0$, $\phi_2 = 0$, which can be obtained from equations (ii) by eliminating $\dfrac{dx}{dt}$ and $\dfrac{dy}{dt}$) shall satisfy equations (i). That c_1, c_2, c_3, and c_4 *can* be so determined, may be seen as follows: by the solution of equations (i), values of x and y and therefore of $\dfrac{dx}{dt}$ and $\dfrac{dy}{dt}$ can be found as functions of t and constant quantities; if these be substituted in equations (ii) the requisite values of c_1, c_2, c_3 and c_4 will be obtained. For an example of the application of this method, see Boole's *Differential Equations*, Chap. IX. Art. 11.

22. If in equations (1) and (2) of Art. 20 we put $R = 0$, and then integrate them, we obtain

$$\frac{1}{r} = \frac{\mu}{h^2}\{1 + e\cos(\theta_0 - \varpi_0)\} \ldots\ldots\ldots\ldots (3),$$

$$\frac{dr}{dt} = \frac{\mu e}{h}\sin(\theta_0 - \varpi_0) \ldots\ldots\ldots\ldots (4),$$

FORMULÆ FOR CALCULATING THE ELEMENTS. 19

$$r^2 \frac{d\theta_0}{dt} = h \quad \text{.....................} (5),$$

where h, e, ϖ_0 are the constants of integration. Equation (3) indicates motion in an ellipse, of which e is the excentricity, ϖ_0 the longitude of perihelion, and h twice the area described in an unit of time. If the mean distance in this ellipse be denoted by a, we have in addition

$$h^2 = \mu a (1 - e^2) \quad \text{.....................} (6).$$

We shall assume (in accordance with the principles of the method of the Variation of Parameters) the first and second integrals of equations (1) and (2), together with equation (6), to retain the same forms when R is not zero; h, e, ϖ_0, and a being in this case considered variable*.

The values of these elements are to be obtained from the condition that the above integrals shall satisfy equations (1) and (2).

23. If their values as calculated for any given time be substituted in equation (3), it will represent an ellipse having a contact of the first order with the actual orbit, since the values of $\frac{dr}{dt}$ and $\frac{d\theta_0}{dt}$ at the common point will be the same for both curves. It is termed the *instantaneous ellipse*, since the planet may for an infinitely small time be supposed to move in it. Moreover, the velocity and direction of motion of the planet will be the same as if it moved in this ellipse, so that if at any time the disturbing force were to cease, the planet would continue to move in the instantaneous ellipse constructed for that time. This is accordingly sometimes given as the definition of the instantaneous ellipse.

* We shall also for convenience suppose the equation $n^2 a^3 = \mu$ to hold in the disturbed orbit, n being of course considered variable.

24. *To obtain formulæ for calculating the elements of the instantaneous ellipse at any time.*

Suppose the value of c required, where c denotes any one of the elements. From equations (3), (4), and (6) we may, by eliminating the other elements, obtain c as a function of r, θ_0, $\dfrac{dr}{dt}$, and h*: let then

$$c = f(r, \theta_0, r', h),$$

where r' is written for $\dfrac{dr}{dt}$. Differentiating, we have

$$\frac{dc}{dt} = \frac{df}{dr}\frac{dr}{dt} + \frac{df}{d\theta_0}\frac{d\theta_0}{dt} + \frac{df}{dr'}\frac{dr'}{dt} + \frac{df}{dh}\frac{dh}{dt}.$$

Now $\dfrac{dr'}{dt} = \dfrac{d^2r}{dt^2} = r\left(\dfrac{d\theta_0}{dt}\right)^2 - \dfrac{\mu}{r^2} + \dfrac{dR}{dr}$, from equation (1);

$\dfrac{dh}{dt} = \dfrac{d}{dt}\left(r^2\dfrac{d\theta_0}{dt}\right) = \dfrac{dR}{d\theta}$, from equation (2);

therefore $\dfrac{dc}{dt} = \dfrac{df}{dr}\dfrac{dr}{dt} + \dfrac{df}{d\theta_0}\dfrac{d\theta_0}{dt} + \dfrac{df}{dr'}\left\{r\left(\dfrac{d\theta_0}{dt}\right)^2 - \dfrac{\mu}{r^2}\right\}$

$$+ \frac{df}{dr'}\frac{dR}{dr} + \frac{df}{dh}\frac{dR}{d\theta}.$$

But, since by hypothesis, if R were zero and c constant, our assumed integrals would still satisfy the differential equations, we have (making R zero and c constant)

$$0 = \frac{df}{dr}\frac{dr}{dt} + \frac{df}{d\theta_0}\frac{d\theta_0}{dt} + \frac{df}{dr'}\left\{r\left(\frac{d\theta_0}{dt}\right)^2 - \frac{\mu}{r^2}\right\};$$

therefore $\dfrac{dc}{dt} = \dfrac{df}{dr'}\dfrac{dR}{dr} + \dfrac{df}{dh}\dfrac{dR}{d\theta}.$

---

* We retain h for convenience, in preference to replacing it by $r^2\dfrac{d\theta_0}{dt}$.

FORMULÆ FOR CALCULATING THE ELEMENTS. 21

Hence in obtaining the formulæ for calculating the elements of the orbit, we may proceed as follows. From equations (3), (4), and (6) we may express the element required as a function of r, θ_0, $\frac{dr}{dt}$, and h. We may then, by differentiating the resulting equation with respect to t as if r and θ_0 were constants, writing $\frac{dR}{dr}$ for $\frac{d^2r}{dt^2}$, and $\frac{dR}{d\theta}$ for $\frac{dh}{dt}$, and eliminating, if necessary, $\frac{dr}{dt}$ and h by means of equations (4) and (6), obtain the differential coefficient of the element required in terms of the elements, the co-ordinates of the planet, and the disturbing force. The result, however, will in every case admit of being expressed in terms of the elements and of the differential coefficients of R with respect to them.

25. *To obtain a formula for calculating the mean distance.*

From equations (3), (4) and (6), if e and ϖ_0 be eliminated, we shall find

$$-\frac{\mu}{a} = \left(\frac{dr}{dt}\right)^2 + \frac{h^2}{r^2} - \frac{2\mu}{r}.$$

Differentiating as if r were constant, and writing $\frac{dR}{dr}$ for $\frac{d^2r}{dt^2}$, $\frac{dR}{d\theta}$ for $\frac{dh}{dt}$, we have

$$\frac{\mu}{a^2}\frac{da}{dt} = 2\left(\frac{dR}{dr}\frac{dr}{dt} + \frac{dR}{d\theta}\frac{d\theta_0}{dt}\right).$$

Now since
$$r = f(nt + \epsilon - \varpi),$$
$$\theta_0 - \varpi_0 = \theta - \varpi = \phi(nt + \epsilon - \varpi),$$

and the forms of $\frac{dr}{dt}$ and $\frac{d\theta_0}{dt}$ are the same as if the elements were constant, we have

$$\frac{dr}{dt} = nf'(nt+\epsilon-\varpi) = n\frac{dr}{d\epsilon},$$

and similarly,
$$\frac{d\theta_0}{dt} = n\frac{d\theta_0}{d\epsilon} = n\frac{d\theta}{d\epsilon};$$

therefore
$$\frac{\mu}{a^2}\frac{da}{dt} = 2n\left(\frac{dR}{dr}\frac{dr}{d\epsilon} + \frac{dR}{d\theta}\frac{d\theta}{d\epsilon}\right)$$

$$= 2n\frac{dR}{d\epsilon},$$

or
$$\frac{da}{dt} = \frac{2na^2}{\mu}\frac{dR}{d\epsilon}.$$

26. This formula may also be obtained as follows. If s denote an arc of the actual path of the planet measured from some fixed point to its position at time t, we have the equation of motion

$$\frac{d^2s}{dt^2} = -\frac{\mu}{r^2}\frac{dr}{ds} + \frac{dR}{ds},$$

and by a known formula

$$\left(\frac{ds}{dt}\right)^2 = \frac{2\mu}{r} - \frac{\mu}{a}.$$

Differentiating the latter we obtain

$$2\frac{ds}{dt}\frac{d^2s}{dt^2} = -\frac{2\mu}{r^2}\frac{dr}{dt} + \frac{\mu}{a^2}\frac{da}{dt},$$

and, multiplying the former by $2\frac{ds}{dt}$,

$$2\frac{ds}{dt}\frac{d^2s}{dt^2} = -\frac{2\mu}{r^2}\frac{dr}{dt} + 2\frac{dR}{ds}\frac{ds}{dt};$$

therefore
$$\frac{\mu}{a^2}\frac{da}{dt} = 2\frac{dR}{ds}\frac{ds}{dt}$$

$$= 2\frac{d(R)}{dt},$$

where $\frac{d(R)}{dt}$ denotes the differential coefficient of R with respect to t, only so far as it involves t through involving the

FORMULÆ FOR CALCULATING THE ELEMENTS. 23

co-ordinates and elements of the *disturbed* planet. Now supposing R expressed as a function of t and the elements (Art. 13), the form of $\dfrac{d(R)}{dt}$ will be the same as if the elements were invariable (see Art. 35): since then t is always coupled with ϵ in the expression $nt + \epsilon$, we have

$$\frac{d(R)}{dt} = n\frac{dR}{d(nt+\epsilon)} = n\frac{dR}{d\epsilon};$$

therefore
$$\frac{da}{dt} = \frac{2na^2}{\mu}\frac{dR}{d\epsilon}.$$

27. *To obtain a formula for calculating the excentricity.*

From equations (3) and (4), if ϖ_0 be eliminated, we obtain

$$\left(\frac{dr}{dt}\right)^2 = \frac{\mu^2 e^2}{h^2} - \left(\frac{h}{r} - \frac{\mu}{h}\right)^2.$$

Differentiating as if r were constant, and writing $\dfrac{dR}{dr}$ for $\dfrac{d^2r}{dt^2}$,

$$\frac{dR}{dr}\frac{dr}{dt} = \frac{\mu^2 e}{h^2}\frac{de}{dt} - \left\{\frac{\mu^2 e^2}{h^3} + \left(\frac{1}{r} + \frac{\mu}{h^2}\right)\left(\frac{h}{r} - \frac{\mu}{h}\right)\right\}\frac{dh}{dt}$$

$$= \frac{\mu^2 e}{h^2}\frac{de}{dt} + \frac{\mu^2(1-e^2)}{h^3}\frac{dR}{d\theta} - \frac{h}{r^2}\frac{dR}{d\theta},$$

since $\dfrac{dh}{dt} = \dfrac{dR}{d\theta}$;

therefore $\dfrac{\mu^2 e}{h^2}\dfrac{de}{dt} = \dfrac{dR}{dr}\dfrac{dr}{dt} + \dfrac{dR}{d\theta}\dfrac{d\theta_0}{dt} - \dfrac{\mu^2(1-e^2)}{h^3}\dfrac{dR}{d\theta}$

$$= n\left(\frac{dR}{dr}\frac{dr}{d\epsilon} + \frac{dR}{d\theta}\frac{d\theta}{d\epsilon}\right) - \frac{\mu^2(1-e^2)}{h^3}\frac{dR}{d\theta},$$

therefore $\dfrac{de}{dt} = \dfrac{nh^2}{\mu^2 e}\dfrac{dR}{d\epsilon} - \dfrac{1-e^2}{he}\left(\dfrac{dR}{d\epsilon} + \dfrac{dR}{d\varpi}\right),$ (Art. 15.)

$$= \frac{na(1-e^2)}{\mu e}\frac{dR}{d\epsilon} - \frac{na\sqrt{(1-e^2)}}{\mu e}\left(\frac{dR}{d\epsilon} + \frac{dR}{d\varpi}\right),$$

since $h^2 = \mu a(1-e^2)$, and $n^2 a^3 = \mu$.

28. This formula may also be deduced from that of the mean distance, by means of the equation
$$h^2 = \mu a (1 - e^2).$$

We have $2h \dfrac{dh}{dt} = \mu (1 - e^2) \dfrac{da}{dt} - 2\mu a e \dfrac{de}{dt}$,

or $2na^2 \sqrt{(1 - e^2)} \dfrac{dR}{d\theta} = 2na^2 (1 - e^2) \dfrac{dR}{d\epsilon} - 2\mu a e \dfrac{de}{dt}$;

therefore $\dfrac{de}{dt} = \dfrac{na(1-e^2)}{\mu e} \dfrac{dR}{d\epsilon} - \dfrac{na\sqrt{(1-e^2)}}{\mu e} \left(\dfrac{dR}{d\epsilon} + \dfrac{dR}{d\varpi} \right).$

29. *To obtain a formula for calculating the longitude of perihelion.*

From equations (3) and (4), if e be eliminated, we obtain
$$\dfrac{dr}{dt} \cot (\theta_0 - \varpi_0) = \dfrac{h}{r} - \dfrac{\mu}{h}.$$

Differentiating as if r and θ_0 were constant, and writing $\dfrac{dR}{dr}$ for $\dfrac{d^2r}{dt^2}$, we obtain

$$\dfrac{dr}{dt} \operatorname{cosec}^2 (\theta_0 - \varpi_0) \dfrac{d\varpi_0}{dt} + \cot (\theta_0 - \varpi_0) \dfrac{dR}{dr} = \left(\dfrac{1}{r} + \dfrac{\mu}{h^2} \right) \dfrac{dh}{dt}$$
$$= \left(\dfrac{1}{r} + \dfrac{\mu}{h^2} \right) \dfrac{dR}{d\theta};$$

therefore $\dfrac{dr}{dt} \operatorname{cosec} (\theta_0 - \varpi_0) \dfrac{d\varpi_0}{dt}$

$$= - \cos (\theta_0 - \varpi_0) \dfrac{dR}{dr} + \left(\dfrac{1}{r} + \dfrac{\mu}{h^2} \right) \sin (\theta_0 - \varpi_0) \dfrac{dR}{d\theta}$$
$$= \dfrac{1}{a} \dfrac{dR}{d\epsilon}; \quad (\text{Art. 17.})$$

but from equation (4) $\dfrac{dr}{dt} \operatorname{cosec} (\theta_0 - \varpi_0) = \dfrac{\mu e}{h}$;

therefore $\dfrac{d\varpi_0}{dt} = \dfrac{h}{\mu e a} \dfrac{dR}{d\epsilon} = \dfrac{na \sqrt{(1 - e^2)} dR}{\mu e}\dfrac{}{d\epsilon}.$

Now if ϖ denote the longitude of perihelion measured on the plane of reference as far as the node, and thence on the plane of the orbit, Ω the longitude of the node on the plane of reference, Ω_0 its longitude on that of the orbit, we have

$$\varpi - \varpi_0 = \Omega - \Omega_0;$$

therefore
$$\frac{d\varpi}{dt} = \frac{d\varpi_0}{dt} + \frac{d\Omega}{dt} - \frac{d\Omega_0}{dt}.$$

Now $\frac{d\Omega}{dt}$ is the angular velocity of the line of nodes on the plane of reference, $\frac{d\Omega_0}{dt}$ its angular velocity on the plane of the orbit;

therefore
$$\frac{d\Omega_0}{dt} = \frac{d\Omega}{dt} \cos i;$$

therefore
$$\frac{d\varpi}{dt} = \frac{d\varpi_0}{dt} + (1 - \cos i) \frac{d\Omega}{dt};$$

or, substituting for $\frac{d\Omega}{dt}$ from Art. 31 or 35,

$$\frac{d\varpi}{dt} = \frac{na \sqrt{(1-e^2)}}{\mu e} \frac{dR}{de} + \frac{na \tan \frac{i}{2}}{\mu \sqrt{(1-e^2)}} \frac{dR}{di}.$$

To obtain formulæ for calculating the longitude of the node, and the inclination.

30. We now return to our third equation of motion,

$$r\phi_1 \frac{d\theta_0}{dt} = Z,$$

or, as it may be written,

$$h\phi_1 = Zr \quad \dots\dots\dots\dots\dots\dots (1).$$

We have seen (Art. 18), that $\phi_2 = 0$; hence the motion of the plane of xy, which coincides with the plane of the orbit, is compounded of the angular velocities ϕ_1 about the axis of x, and ϕ_3 about the axis of z. Now the former is equivalent to an angular velocity $\phi_1 \cos(\theta - \Omega)$ about the line of nodes, and an angular velocity $\phi_1 \sin(\theta - \Omega)$ about an axis perpendicular to it in the plane of the orbit: but the angular velocities of the plane of the orbit about these axes are $\dfrac{di}{dt}$ and $\sin i \dfrac{d\Omega}{dt}$ respectively; therefore

$$\phi_1 \cos(\theta - \Omega) = \frac{di}{dt}, \quad \phi_1 \sin(\theta - \Omega) = \sin i \frac{d\Omega}{dt}.$$

Hence, by equation (1),

$$h \frac{di}{dt} = Zr \cos(\theta - \Omega) \dots\dots\dots\dots (2),$$

$$h \sin i \frac{d\Omega}{dt} = Zr \sin(\theta - \Omega) \dots\dots\dots\dots (3).$$

31. In order to determine Z we must suppose the curve of reference perpendicular to the plane of the orbit. If we denote by s an arc of this curve measured from some fixed point up to the planet, we have by Art. 8, $Z = \dfrac{dR}{ds}$. Now the position of a point on the curve of reference may be determined by its polar co-ordinates r, θ on a plane passing through it and the sun, the inclination i of this plane to the plane of reference, and the longitude Ω of its node. Since, however, an infinite number of planes can be drawn through two given points, we must introduce some further condition to fix the position of that on which r and θ are measured.

First, then, let it pass through SN the line of the nodes: let P be the position of the planet, and suppose the curve of reference a circle AP with its centre C in SN. Through SN

FORMULÆ FOR CALCULATING THE ELEMENTS.

draw a plane inclined at a small angle δi to the plane of the orbit, cutting the circle AP in p, and let $Pp = \delta s$; then

$$Pp = CP . \delta i,$$

or

$$\delta s = r \sin (\theta - \Omega) \delta i;$$

therefore

$$\frac{di}{ds} = \frac{1}{r \sin (\theta - \Omega)}.$$

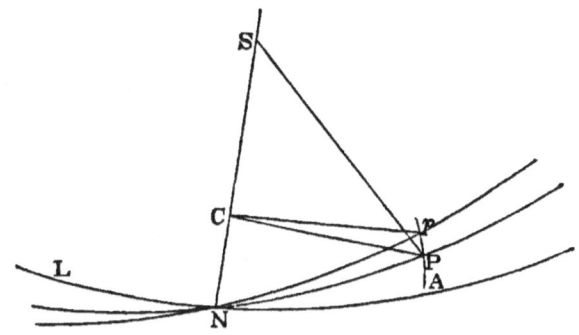

Now, (see Art. 13), R may be expressed in the form

$$R = f(r, \theta, \Omega, i),$$

in which, if the series for r and θ be substituted, R will be expressed as a function of t and the elements. Hence

$$\frac{dR}{ds} = \frac{dR}{dr}\frac{dr}{ds} + \frac{dR}{d\theta}\frac{d\theta}{ds} + \frac{dR}{d\Omega}\frac{d\Omega}{ds} + \frac{dR}{di}\frac{di}{ds},$$

and

$$\frac{dr}{ds} = 0, \quad \frac{d\theta}{ds} = 0, \quad \frac{d\Omega}{ds} = 0, \quad \frac{di}{ds} = \frac{1}{r \sin (\theta - \Omega)};$$

therefore

$$\frac{dR}{ds} = \frac{1}{r \sin (\theta - \Omega)} \frac{dR}{di}.$$

On substituting this value for Z in equation (3), we obtain

$$h \sin i \frac{d\Omega}{dt} = \frac{dR}{di},$$

or since $h^2 = \mu a (1 - e^2)$, and $n^2 a^3 = \mu$,

$$\frac{d\Omega}{dt} = \frac{na}{\mu \sqrt{(1-e^2)} \sin i} \cdot \frac{dR}{di}.$$

32. Again, suppose the plane on which r and θ are measured, instead of passing through SN, to pass through a line SC in the plane of the orbit perpendicular to SN; and take for the curve of reference a circle AP with its centre C in SC. Through SC draw a plane inclined at a small angle to the plane of the orbit, cutting the sphere in the great circle nmp, and the circle AP in p. Draw Nm perpendicular to np.

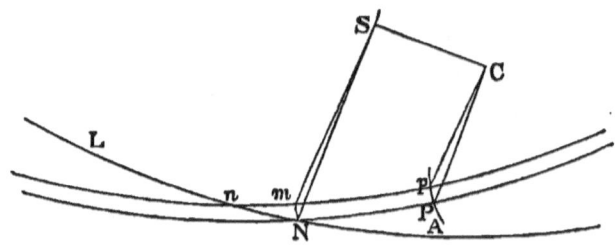

Then Nm which measures the inclination of the two planes $= -\delta\Omega \sin i$, and

$$CP = r \cos(\theta - \Omega);$$

hence if $Pp = \delta s$,

$$\delta s = -r \cos(\theta - \Omega) \delta\Omega \sin i;$$

therefore

$$\frac{d\Omega}{ds} = \frac{-1}{r \cos(\theta - \Omega) \sin i}.$$

Again, $\quad \delta\theta = Ln + np - (LN + NP)$

$\quad\quad\quad\quad = nm - nN$

$\quad\quad\quad\quad = -\delta\Omega \cos i + \delta\Omega;$

therefore $\dfrac{d\theta}{ds} = (1 - \cos i)\dfrac{d\dot\Omega}{ds};$

also $\dfrac{dr}{ds} = 0, \; \dfrac{di}{ds} = 0.$

Hence $\dfrac{dR}{ds} = \dfrac{dR}{d\theta}\dfrac{d\theta}{ds} + \dfrac{dR}{d\Omega}\dfrac{d\Omega}{ds}$

$= -\left\{\dfrac{dR}{d\Omega} + (1 - \cos i)\dfrac{dR}{d\theta}\right\}\dfrac{1}{r\cos(\theta - \Omega)\sin i}.$

On substituting this value for Z in equation (2), we have

$\dfrac{di}{dt} = -\dfrac{1}{h\sin i}\left\{\dfrac{dR}{d\Omega} + (1 - \cos i)\dfrac{dR}{d\theta}\right\}$

$= -\dfrac{1}{h\sin i}\left\{\dfrac{dR}{d\Omega} + 2\sin^2\dfrac{i}{2}\left(\dfrac{dR}{de} + \dfrac{dR}{d\varpi}\right)\right\}$ (Art. 15),

$= -\dfrac{na}{\mu\sqrt{(1-e^2)}}\left\{\dfrac{1}{\sin i}\dfrac{dR}{d\Omega} + \tan\dfrac{i}{2}\left(\dfrac{dR}{de} + \dfrac{dR}{d\varpi}\right)\right\}.$

33. The only remaining element is the epoch, but before proceeding to obtain a formula for its calculation, we shall give another method of obtaining the results of the last two articles.

34. *To obtain a formula for calculating the inclination.* (*Second method.*)

If the motion of the planet be referred to the polar co-ordinates of its projection on the fixed plane of reference and its distance from this plane, we have the equation

$\dfrac{1}{r_1}\dfrac{d}{dt}\left(r_1^2\dfrac{d\theta_1}{dt}\right) = \dfrac{1}{r_1}\dfrac{dR}{d\theta_1}$ (See Art. 9),

or $\dfrac{d}{dt}\left(r_1^2\dfrac{d\theta_1}{dt}\right) = \dfrac{dR}{d\theta_1}.$

Now if $\delta A, \; \delta A_1$ denote the vectorial areas swept out in the time δt on the plane of the orbit and the plane of reference respectively, we have

$$\delta A_1 = \delta A \cos i:$$

but $\quad \delta A_1 = \frac{1}{2} r_1^2 \delta \theta_1, \quad \delta A = \frac{1}{2} r^2 \delta \theta;$

therefore $\quad r_1^2 \frac{d\theta_1}{dt} = r^2 \frac{d\theta}{dt} \cos i = h \cos i.$

Hence our equation of motion becomes

$$\frac{d}{dt}(h \cos i) = \frac{dR}{d\theta_1},$$

or $\quad -h \sin i \frac{di}{dt} + \cos i \frac{dh}{dt} = \frac{dR}{d\theta_1};$

therefore $-h \sin i \frac{di}{dt} = \frac{dR}{d\theta_1} - \cos i \frac{dh}{dt}$

$$= \frac{dR}{d\theta_1} - \cos i \frac{dR}{d\theta}$$

$$= \frac{dR}{d\Omega} + (1 - \cos i) \frac{dR}{d\theta}, \quad \text{(Art. 16)},$$

$$= \frac{dR}{d\Omega} + (1 - \cos i) \left(\frac{dR}{d\epsilon} + \frac{dR}{d\varpi}\right), \quad \text{(Art. 15)};$$

therefore $\dfrac{di}{dt} = -\dfrac{1}{h \sin i} \left\{\dfrac{dR}{d\Omega} + (1 - \cos i) \left(\dfrac{dR}{d\epsilon} + \dfrac{dR}{d\varpi}\right)\right\}$

$$= -\frac{na}{\mu \sqrt{(1-e^2)}} \left\{\frac{1}{\sin i} \frac{dR}{d\Omega} + \tan \frac{i}{2} \left(\frac{dR}{d\epsilon} + \frac{dR}{d\varpi}\right)\right\}.$$

35. *To obtain a formula for calculating the longitude of the node.* (*Second method.*)

Since the velocity of the planet at any time can be expressed in terms of the co-ordinates and elements of the instantaneous ellipse constructed for that time, in the same form as

FORMULÆ FOR CALCULATING THE ELEMENTS. 31

if it moved in this ellipse, its component in any direction can also be so expressed. Hence, considering R as a function of r_1, θ_1, z (see Art. 11), the values of $\frac{dr_1}{dt}$, $\frac{d\theta_1}{dt}$, $\frac{dz}{dt}$, and therefore of $\frac{d(R)}{dt}$ (where $\frac{d(R)}{dt}$ denotes that R is to be differentiated with respect to t only so far as it involves t through involving the co-ordinates of the disturbed planet) may be expressed in the same forms as if the elements were invariable.

Now we have seen (Art. 13) that R may be thus expressed:
$$R = f(r, \theta, \Omega, i);$$
or since evidently, $\theta - \Omega = \theta_0 - \Omega_0$,
$$R = f(r, \theta_0 + \Omega - \Omega_0, \Omega, i)^*.$$

We have then, considering the elements variable,
$$\frac{d(R)}{dt} = \frac{dR}{dr}\frac{dr}{dt} + \frac{dR}{d\theta_0}\frac{d(\theta_0 + \Omega - \Omega_0)}{dt} + \frac{dR}{d\Omega}\frac{d\Omega}{dt} + \frac{dR}{di}\frac{di}{dt},$$
and, considering them invariable,
$$\frac{d(R)}{dt} = \frac{dR}{dr}\frac{dr}{dt} + \frac{dR}{d\theta_0}\frac{d\theta_0}{dt}.$$

Equating the two values of $\frac{d(R)}{dt}$, we obtain
$$\frac{dR}{d\theta_0}\left(\frac{d\Omega}{dt} - \frac{d\Omega_0}{dt}\right) + \frac{dR}{d\Omega}\frac{d\Omega}{dt} + \frac{dR}{di}\frac{di}{dt} = 0.$$

Now
$$\frac{dR}{d\theta_0} = \frac{dR}{d\theta} = \frac{dR}{d\epsilon} + \frac{dR}{d\varpi},$$

* We have made this transformation, because, although the value of $\frac{d\theta_0}{dt}$ is the same in form as if the elements were invariable, this is not the case with $\frac{d\theta}{dt}$.

and (see Art. 29), $\dfrac{d\Omega_{\text{o}}}{dt} = \dfrac{d\Omega}{dt} \cos i$;

therefore $\left\{ \dfrac{dR}{d\Omega} + (1 - \cos i) \left(\dfrac{dR}{d\epsilon} + \dfrac{dR}{d\varpi} \right) \right\} \dfrac{d\Omega}{dt} + \dfrac{dR}{di} \dfrac{di}{dt} = 0$.

Substituting for $\dfrac{di}{dt}$ its value

$$\dfrac{d\Omega}{dt} = \dfrac{na}{\mu \sqrt{(1 - e^2)} \sin i} \dfrac{dR}{di}.$$

36. *To obtain a formula for calculating the longitude of the epoch.*

If R be expressed in terms of t and the elements (see Art. 13), since $nt + \epsilon$ always occurs as one symbol, we may write

$$R = f(nt + \epsilon,\ a,\ e,\ \varpi,\ \Omega,\ i).$$

Differentiating, the elements being considered variable, we have

$$\dfrac{d(R)}{dt} = \dfrac{dR}{d\epsilon} \dfrac{d(nt+\epsilon)}{dt} + \dfrac{dR}{da} \dfrac{da}{dt} + \dfrac{dR}{de} \dfrac{de}{dt} + \dfrac{dR}{d\varpi} \dfrac{d\varpi}{dt}$$
$$+ \dfrac{dR}{d\Omega} \dfrac{d\Omega}{dt} + \dfrac{dR}{di} \dfrac{di}{dt},$$

and differentiating as if the elements were invariable, which is permissible for the reason explained in the last Article,

$$\dfrac{d(R)}{dt} = n \dfrac{dR}{d\epsilon}.$$

Equating the two values of $\dfrac{d(R)}{dt}$,

$$n \dfrac{dR}{d\epsilon} = \dfrac{dR}{d\epsilon} \left(n + t \dfrac{dn}{dt} + \dfrac{d\epsilon}{dt} \right) + \dfrac{dR}{da} \dfrac{da}{dt} + \dfrac{dR}{de} \dfrac{de}{dt} + \dfrac{dR}{d\varpi} \dfrac{d\varpi}{dt}$$
$$+ \dfrac{dR}{d\Omega} \dfrac{d\Omega}{dt} + \dfrac{dR}{di} \dfrac{di}{dt}.$$

FORMULÆ FOR CALCULATING THE ELEMENTS.

Substituting for $\dfrac{da}{dt}$, $\dfrac{d\epsilon}{dt}$, &c., their values

$$0 = \frac{dR}{d\epsilon}\left(t\frac{dn}{dt} + \frac{d\epsilon}{dt}\right) + \frac{2na^2}{\mu}\frac{dR}{d\epsilon}\frac{dR}{da} + \frac{na(1-e^2)}{\mu e}\frac{dR}{d\epsilon}\frac{dR}{de}$$

$$-\frac{na\sqrt{(1-e^2)}}{\mu e}\left(\frac{dR}{d\epsilon} + \frac{dR}{d\varpi}\right)\frac{dR}{de} + \frac{na\sqrt{(1-e^2)}}{\mu e}\frac{dR}{d\epsilon}\frac{dR}{d\varpi}$$

$$+\frac{na\tan\frac{i}{2}}{\mu\sqrt{(1-e^2)}}\frac{dR}{di}\frac{dR}{d\varpi} + \frac{na}{\mu\sqrt{(1-e^2)}\sin i}\frac{dR}{di}\frac{dR}{d\Omega}$$

$$-\frac{na}{\mu\sqrt{(1-e^2)}\sin i}\frac{dR}{d\Omega}\frac{dR}{di} - \frac{na\tan\frac{i}{2}}{\mu\sqrt{(1-e^2)}}\left(\frac{dR}{d\epsilon} + \frac{dR}{d\varpi}\right)\frac{dR}{di},$$

$$= \frac{dR}{d\epsilon}\left(t\frac{dn}{dt} + \frac{d\epsilon}{dt}\right) + \frac{2na^2}{\mu}\frac{dR}{d\epsilon}\frac{dR}{da}$$

$$-\frac{na\sqrt{(1-e^2)}}{\mu e}\{1-\sqrt{(1-e^2)}\}\frac{dR}{d\epsilon}\frac{dR}{de}$$

$$-\frac{na\tan\frac{i}{2}}{\mu\sqrt{(1-e^2)}}\frac{dR}{d\epsilon}\frac{dR}{di}.$$

Dividing every term by $\dfrac{dR}{d\epsilon}$, and transposing, we obtain

$$\frac{d\epsilon}{dt} = -t\frac{dn}{dt} - \frac{2na^2}{\mu}\frac{dR}{da} + \frac{na\sqrt{(1-e^2)}}{\mu e}\{1-\sqrt{(1-e^2)}\}\frac{dR}{de}$$

$$+\frac{na\tan\frac{i}{2}}{\mu\sqrt{(1-e^2)}}\frac{dR}{di}.$$

37. Of the formulæ which have been obtained for calculating the elements of the orbit, that of the preceding article is the only one which contains a term proportional to the

time*. It may, however, be replaced by one in which no such term exists. For, let ξ denote the mean longitude, then
$$\xi = nt + \epsilon;$$

therefore
$$\frac{d\xi}{dt} = n + t\frac{dn}{dt} + \frac{d\epsilon}{dt},$$

or
$$\frac{d}{dt}(\xi - \int n dt) = t\frac{dn}{dt} + \frac{d\epsilon}{dt}.$$

Now let
$$\xi = \int n dt + \epsilon',$$
then by the formula of the last article
$$\frac{d\epsilon'}{dt} = -\frac{2na^2}{\mu}\frac{dR}{da} + \frac{na\sqrt{(1-e^2)}}{\mu e}\{1 - \sqrt{(1-e^2)}\}\frac{dR}{de}$$
$$+ \frac{na\tan\frac{i}{2}}{\mu\sqrt{(1-e^2)}}\frac{dR}{di}.$$

Since in the elliptic formulæ ϵ never occurs except when coupled with nt, in the expression $nt + \epsilon$, it will be altogether eliminated if for $nt + \epsilon$ we write $\int n dt + \epsilon'$.

Considered as replacing the element ϵ, ϵ' is called the *epoch**. Since however we shall never have occasion to employ the formula of the preceding article, the accent will in future be omitted.

$\int n dt$ is termed the *mean motion* in the disturbed orbit, and is denoted by ζ.

38. To obtain a formula for calculating ζ, we have
$$\frac{d^2\zeta}{dt^2} = \frac{dn}{dt},$$

---

* The reader, if acquainted with the Lunar Theory, will have already seen the inconvenience of such terms.

† It may be noticed that if n were constant, ϵ' would be identical with ϵ.

FORMULÆ FOR CALCULATING THE ELEMENTS.

and differentiating the equation $n^2 a^3 = \mu$

$$2na^3 \frac{dn}{dt} + 3n^2 a^2 \frac{da}{dt} = 0;$$

therefore
$$\frac{dn}{dt} = -\frac{3n}{2a}\frac{da}{dt} = -\frac{3n^2 a}{\mu}\frac{dR}{d\epsilon},$$

or $\quad \dfrac{d^2\zeta}{dt^2} = -\dfrac{3n^2 a}{\mu}\dfrac{dR}{d\epsilon}.$

39. We will here recapitulate the formulæ which have been obtained for calculating the elements of the orbit.

(i) $\quad \dfrac{da}{dt} = \dfrac{2na^2}{\mu}\dfrac{dR}{d\epsilon}.$

(ii) $\quad \dfrac{de}{dt} = \dfrac{na(1-e^2)}{\mu e}\dfrac{dR}{d\epsilon} - \dfrac{na\sqrt{(1-e^2)}}{\mu e}\left(\dfrac{dR}{d\epsilon} + \dfrac{dR}{d\varpi}\right).$

(iii) $\quad \dfrac{d\varpi}{dt} = \dfrac{na\sqrt{(1-e^2)}}{\mu e}\dfrac{dR}{d\epsilon} + \dfrac{na\tan\frac{i}{2}}{\mu\sqrt{(1-e^2)}}\dfrac{dR}{di}.$

(iv) $\quad \dfrac{d\epsilon}{dt} = -\dfrac{2na^2}{\mu}\dfrac{dR}{da} + \dfrac{na\sqrt{(1-e^2)}}{\mu e}\{1-\sqrt{(1-e^2)}\}\dfrac{dR}{de}$

$$+ \dfrac{na\tan\frac{i}{2}}{\mu\sqrt{(1-e^2)}}\dfrac{dR}{di}.$$

(v) $\quad \dfrac{d\Omega}{dt} = \dfrac{na}{\mu(1-e^2)\sin i}\dfrac{dR}{di}.$

(vi) $\quad \dfrac{di}{dt} = -\dfrac{na}{\mu\sqrt{(1-e^2)}}\left\{\dfrac{1}{\sin i}\dfrac{dR}{d\Omega} + \tan\dfrac{i}{2}\left(\dfrac{dR}{d\epsilon} + \dfrac{dR}{d\varpi}\right)\right\}.$

We have also (vii) the equation

$$\frac{d^2\zeta}{dt^2} = -\frac{3n^2 a}{\mu}\frac{dR}{d\epsilon},$$

3—2

but this forms no new relation, since it has been deduced from (i).

40. When the elements have been calculated by means of the above formulæ, the position of the planet will be given by the equations

$$r = a\left\{1 + \frac{1}{2}e^2 - e\cos(\zeta + \epsilon - \varpi) - \frac{1}{2}e^2\cos 2(\zeta + \epsilon - \varpi) - \ldots\right\},$$

$$\theta = \zeta + \epsilon + 2e\sin(\zeta + \epsilon - \varpi) + \frac{5}{4}e^2\cos 2(\zeta + \epsilon - \varpi) + \ldots.$$

CHAPTER III.

DEVELOPMENT OF THE DISTURBING FUNCTION.

41. In the first chapter we have obtained equations by means of which R may be expressed in terms of the time and the elements of the orbit; we now proceed to shew how the actual development may be effected in a series ascending by powers and products of the excentricities and the tangents of the inclinations. In the Planetary Theory these are extremely small, and the series will converge rapidly. Accordingly in the present treatise small quantities of orders higher than the second will be neglected*.

42. If we recur to Art. 11, it will be seen that, considering only one disturbing planet,

$$R = m' \left[\frac{1}{\{r_1^2 + r_1'^2 - 2r_1 r_1' \cos(\theta_1 - \theta_1') + (z - z')^2\}^{\frac{1}{2}}} - \frac{r_1 r_1' \cos(\theta_1 - \theta_1') + zz'}{(r_1'^2 + z'^2)^{\frac{3}{2}}} \right].$$

The first step towards the required development will be the expansion of r_1, r_1', θ_1, θ_1', z and z' in terms of the time

---

* We may remark that to this order of approximation the inclinations, their sines, and tangents will be equal.

and the elements of the orbit. For this purpose we may employ the equations which have already been obtained in Art. 13, viz.:

$$\tan(\theta_1 - \Omega) = \cos i \tan(\theta - \Omega),$$
$$\sin \lambda = \sin i \sin(\theta - \Omega),$$
$$r_1 = r \cos \lambda, \qquad z = r \sin \lambda,$$
$$r = a\left\{1 + \frac{1}{2}e^2 - e\cos(nt + \epsilon - \varpi) - \frac{1}{2}e^2\cos 2(n + \epsilon - \varpi) - \ldots\right\},$$
$$\theta = nt + \epsilon + 2e\sin(nt + \epsilon - \varpi) + \frac{5}{4}e^2\sin 2(nt + \epsilon - \varpi) + \ldots,$$

with similar equations involving the co-ordinates and elements of the disturbing planet.

(i) To expand r_1. We have

$$r_1 = r\cos\lambda = r(1 - \sin^2\lambda)^{\frac{1}{2}}$$
$$= r\left(1 - \frac{1}{2}\sin^2\lambda + \ldots\right)$$
$$= r\left\{1 - \frac{1}{2}\sin^2 i \sin^2(\theta - \Omega) + \ldots\right\}$$
$$= r\left\{1 - \frac{1}{2}\tan^2 i \sin^2(\theta - \Omega) + \ldots\right\}$$

to the same order of approximation,

$$= r\left\{1 - \frac{1}{4}\tan^2 i + \frac{1}{4}\tan^2 i \cos 2(\theta - \Omega) + \ldots\right\};$$

or substituting the expansions for r and θ,

$$r_1 = a\left\{1 + \frac{1}{2}e^2 - \frac{1}{4}\tan^2 i - e\cos(nt + \epsilon - \varpi) - \frac{1}{2}e^2\cos 2(nt + \epsilon - \varpi)\right.$$
$$\left. + \frac{1}{4}\tan^2 i \cos 2(nt + \epsilon - \Omega) + \ldots\right\}$$
$$= a(1 + u), \text{ suppose.}$$

Similarly, $\quad r_1' = a'(1+u')$.

(ii) To expand θ_1. We have

$$\tan(\theta_1 - \theta) = \tan\{(\theta_1 - \Omega) - (\theta - \Omega)\}$$

$$= \frac{\tan(\theta_1 - \Omega) - \tan(\theta - \Omega)}{1 + \tan(\theta_1 - \Omega)\tan(\theta - \Omega)}$$

$$= \frac{(\cos i - 1)\tan(\theta - \Omega)}{1 + \cos i \tan^2(\theta - \Omega)}$$

$$= \frac{-2\sin^2\frac{i}{2}\tan(\theta - \Omega)}{1 + \tan^2(\theta - \Omega) - 2\sin^2\frac{i}{2}\tan^2(\theta - \Omega)}$$

$$= \frac{-\sin^2\frac{i}{2}\sin 2(\theta - \Omega)}{1 - 2\sin^2\frac{i}{2}\sin^2(\theta - \Omega)}$$

$$= -\sin^2\frac{i}{2}\sin 2(\theta - \Omega) - \ldots ;$$

therefore $\quad \theta_1 - \theta = -\sin^2\frac{i}{2}\sin 2(\theta - \Omega) - \ldots$

$$= -\tan^2\frac{i}{2}\sin 2(\theta - \Omega) - \ldots,$$

to the same order of approximation; or, substituting the expansion for θ,

$$\theta_1 = nt + \epsilon + 2e\sin(nt + \epsilon - \varpi) + \frac{5}{4}e^2\sin 2(nt + \epsilon - \varpi)$$

$$- \tan^2\frac{i}{2}\sin 2(nt + \epsilon - \Omega) + \ldots$$

$= nt + \epsilon + v$, suppose.

Similarly, $\quad \theta_1' = n't + \epsilon' + v'$.

(iii) To expand z. We have
$$z = r \sin \lambda = r \sin i \sin (\theta - \Omega)$$
$$= r \tan i \sin (\theta - \Omega) - \dots$$

to the second order; or, substituting the expansions for r and θ,
$$z = a \{\tan i \sin (nt + \epsilon - \Omega) + \dots\}.$$

A similar expression may be found for z'.

43. Having obtained the expansions of $r_{,}$, $r_{,}'$, $\theta_{,}$, $\theta_{,}'$, z, z' we must now substitute them in the expression for R. This may be effected as follows.

Let R' be the value of R when u, u', v, v' are severally zero: then, writing ϕ for $nt + \epsilon - (n't + \epsilon')$ we have

$$R' = m' \left[\{a^2 + a'^2 - 2aa' \cos \phi + (z - z')^2\}^{-\frac{1}{2}} \right.$$
$$\left. - (aa' \cos \phi + zz') (a'^2 + z'^2)^{-\frac{3}{2}} \right]$$

$$= m' \left[(a^2 + a'^2 - 2aa' \cos \phi)^{-\frac{1}{2}} - \frac{a}{a'^2} \cos \phi \right]$$

$$- m' \left[\frac{1}{2} (a^2 + a'^2 - 2aa' \cos \phi)^{-\frac{3}{2}} (z - z')^2 \right.$$
$$\left. + \frac{1}{a'^3} zz' - \frac{3}{2} \frac{a}{a'^2} z'^2 \cos \phi \right]$$

$+ \dots\dots\dots\dots$

Now $R = R' + \dfrac{dR'}{da} au + \dfrac{dR'}{da'} a'u' + \dfrac{dR'}{d\phi} (v - v')$

$+ \dfrac{1}{2} \left\{ \dfrac{d^2R'}{da^2} a^2 u^2 + \dfrac{d^2R'}{da'^2} a'^2 u'^2 + \dfrac{d^2R'}{d\phi^2} (v - v')^2 \right\}$

$+ \dfrac{d^2R'}{da\, da'} aa'\, uu' + \dfrac{d^2R'}{da\, d\phi} au\, (v - v') + \dfrac{d^2R'}{da'\, d\phi} a'u'\, (v - v')$

$+ \dots\dots\dots\dots$

DEVELOPMENT OF THE DISTURBING FUNCTION. 41

44. It will be shewn in a subsequent article that

$$(a^2 + a'^2 - 2aa' \cos \phi)^{-\frac{1}{2}},$$

can be expanded in a series of the form

$$\frac{1}{2} A_0 + A_1 \cos \phi + A_2 \cos 2\phi + \ldots + A_k \cos k\phi + \ldots$$

Assume then

$$(a^2 + a'^2 - 2aa' \cos \phi)^{-\frac{1}{2}} = \frac{1}{2} C_0 + C_1 \cos \phi + C_2 \cos 2\phi + \ldots$$

$$(a^2 + a'^2 - 2aa' \cos \phi)^{-\frac{3}{2}} = \frac{1}{2} D_0 + D_1 \cos \phi + D_2 \cos 2\phi + \ldots$$

Thus

$$R = m' \left\{ \frac{1}{2} C_0 + \left(C_1 - \frac{a}{a'^2} \right) \cos \phi + C_2 \cos 2\phi + \ldots \right\}$$

$$+ m'au \left\{ \frac{1}{2} \frac{dC_0}{da} + \left(\frac{dC_1}{da} - \frac{1}{a'^2} \right) \cos \phi + \frac{dC_2}{da} \cos 2\phi + \ldots \right\}$$

$$+ m'a'u' \left\{ \frac{1}{2} \frac{dC_0}{da'} + \left(\frac{dC_1}{da'} + \frac{2a}{a'^3} \right) \cos \phi + \frac{dC_2}{da'} \cos 2\phi + \ldots \right\}$$

$$- m'(v - v') \left\{ \left(C_1 - \frac{a}{a'^2} \right) \sin \phi + 2C_2 \sin 2\phi + \ldots \right\}$$

$$+ \frac{m'a^2u^2}{2} \left\{ \frac{1}{2} \frac{d^2C_0}{da^2} + \frac{d^2C_1}{da^2} \cos \phi + \frac{d^2C_2}{da^2} \cos 2\phi + \ldots \right\}$$

$$+ \frac{m'a'^2u'^2}{2} \left\{ \frac{1}{2} \frac{d^2C_0}{da'^2} + \left(\frac{d^2C_1}{da'^2} - \frac{6a}{a'^4} \right) \cos \phi + \frac{d^2C_2}{da'^2} \cos 2\phi + \ldots \right\}$$

$$- \frac{m'(v - v')^2}{2} \left\{ \left(C_1 - \frac{a}{a'^2} \right) \cos \phi + 4C_2 \cos 2\phi + \ldots \right\}$$

$$+ m'aa'uu' \left\{ \frac{1}{2} \frac{d^2C_0}{da\,da'} + \left(\frac{d^2C_1}{da\,da'} + \frac{2}{a'^3} \right) \cos \phi + \frac{d^2C_2}{da\,da'} \cos 2\phi + \ldots \right\}$$

$$- m'au(v-v')\left\{\left(\frac{dC_1}{da} - \frac{1}{a'^2}\right)\sin\phi + 2\frac{dC_2}{da}\sin 2\phi + \ldots\right\}$$

$$- m'a'u'(v-v')\left\{\left(\frac{dC_1}{da'} + \frac{2a}{a'^3}\right)\sin\phi + 2\frac{dC_2}{da'}\sin 2\phi + \ldots\right\}$$

$$- \frac{m'(z-z')^2}{2}\left\{\frac{1}{2}D_0 + D_1\cos\phi + D_2\cos 2\phi + \ldots\right\}$$

$$- m'\left\{\frac{zz'}{a'^3} - \frac{3}{2}\frac{a}{a'^2}z'^2\cos\phi + \ldots\right\}.$$

45. By Art. 42,
$$u = \frac{1}{2}e^2 - \frac{1}{4}\tan^2 i - e\cos(nt + \epsilon - \varpi) - \frac{1}{2}e^2\cos 2(nt + \epsilon - \varpi)$$
$$+ \frac{1}{4}\tan^2 i \cos 2(nt + \epsilon - \Omega) + \ldots$$

$$v = 2e\sin(nt + \epsilon - \varpi) + \frac{5}{4}e^2\sin 2(nt + \epsilon - \varpi)$$
$$- \tan^2\frac{i}{2}\sin 2(nt + \epsilon - \Omega) + \ldots,$$

$z = a\{\tan i \sin(nt + \epsilon - \Omega) + \ldots\}$,

with similar expressions for u', v', z'.

Hence $u^2 = e^2\cos^2(nt + \epsilon - \varpi) + \ldots$
$$= \frac{e^2}{2} + \frac{e^2}{2}\cos 2(nt + \epsilon - \varpi) + \ldots,$$

$(v-v')^2 = 4e^2\sin^2(nt + \epsilon - \varpi) + 4e'^2\sin^2(n't + \epsilon' - \varpi')$
$$- 8ee'\sin(nt + \epsilon - \varpi)\sin(n't + \epsilon' - \varpi') + \ldots$$
$$= 2(e^2 + e'^2) - 2e^2\cos 2(nt + \epsilon - \varpi) - 2e'^2\cos 2(n't + \epsilon' - \varpi')$$
$$- 4ee'\cos(\phi - \varpi + \varpi') + 4ee'\cos\{(n+n')t + \epsilon + \epsilon' - \varpi - \varpi'\} + \ldots,$$

$uu' = ee'\cos(nt + \epsilon - \varpi)\cos(n't + \epsilon' - \varpi') + \ldots$

$$= \frac{ee'}{2} \cos (\phi - \varpi + \varpi') + \frac{ee'}{2} \cos \{(n+n')t + \epsilon + \epsilon' - \varpi - \varpi'\} + \dots$$

$$u(v-v') = -e^2 \sin 2(nt + \epsilon - \varpi)$$
$$+ 2ee' \cos(nt + \epsilon - \varpi) \sin(n't + \epsilon' - \varpi') + \dots$$
$$= -e^2 \sin 2(nt + \epsilon - \varpi) - ee' \sin(\phi - \varpi + \varpi')$$
$$+ ee' \sin\{(n+n')t + \epsilon + \epsilon' - \varpi - \varpi'\} + \dots,$$

$$(z-z')^2 = \frac{a^2 \tan^2 i}{2} + \frac{a'^2 \tan^2 i'}{2} - \frac{a^2 \tan^2 i}{2} \cos 2(nt + \epsilon - \Omega)$$

$$- \frac{a'^2 \tan^2 i'}{2} \cos 2(n't + \epsilon' - \Omega') - aa' \tan i \tan i' \cos(\phi - \Omega + \Omega')$$
$$+ aa' \tan i \tan i' \cos\{(n+n')t + \epsilon + \epsilon' - \Omega - \Omega'\} + \dots$$

&c.

46. If these values be substituted in Art. 44, it will be seen that cosines will be multiplied only by cosines, and sines by sines. Hence the series will consist of two parts, one independent of t explicitly, and the other consisting of periodical terms of the form

$$P \cos \{(pn \pm qn')t + Q\},$$

where p and q are any positive integers or zero, P is a function of the mean distances, excentricities, and inclinations; and Q a function of the longitudes of perihelia, nodes, and epochs. The former part is denoted by the symbol F: we proceed to determine its value as far as the second order of small quantities.

47. *To determine that part of* R *which is independent of the time explicitly.*

If those terms only be written down which either are, or after reduction will become, independent of t, we have

$$m' \left\{ \frac{C_0}{2} + \frac{a}{2} \frac{dC_0}{da} \left(\frac{e^2}{2} - \frac{\tan^2 i}{4} \right) + \frac{a'}{2} \frac{dC_0}{da'} \left(\frac{e'^2}{2} - \frac{\tan^2 i'}{4} \right) \right.$$

$$+ \frac{a^2}{4} \frac{d^2 C_0}{da^2} \frac{e^2}{2} + \frac{a'^2}{4} \frac{d^2 C_0}{da'^2} \frac{e'^2}{2} - \frac{D_0}{4} \left(\frac{a^2 \tan^2 i}{2} + \frac{a'^2 \tan^2 i''}{2} \right)$$

$$+ \frac{1}{2} \left(C_1 - \frac{a}{a'^2} \right) \cos \phi \, 4ee' \cos (\phi - \varpi + \varpi')$$

$$+ aa' \left(\frac{d^2 C_1}{da \, da'} + \frac{2}{a'^3} \right) \cos \phi \, \frac{ee'}{2} \cos (\phi - \varpi + \varpi')$$

$$+ a \left(\frac{dC_1}{da} - \frac{1}{a'^2} \right) \sin \phi \, ee' \sin (\phi - \varpi + \varpi')$$

$$+ a' \left(\frac{dC_1}{da'} + \frac{2a}{a'^3} \right) \sin \phi \, ee' \sin (\phi - \varpi + \varpi')$$

$$+ \frac{D_1}{2} \cos \phi \, aa' \tan i \tan i'' \cos (\phi - \Omega + \Omega') + \ldots \Big\} .$$

Now $\cos \phi \cos (\phi - \varpi + \varpi')$ and $\sin \phi \sin (\phi - \varpi - \varpi')$ contain the term $\frac{1}{2} \cos (\varpi - \varpi')$, $\cos \phi \cos (\phi - \Omega + \Omega')$ contains the term $\frac{1}{2} \cos (\Omega - \Omega')$; hence

$$F = m' \left\{ \frac{C_0}{2} + \frac{1}{4} \left(a \frac{dC_0}{da} + \frac{a^2}{2} \frac{d^2 C_0}{da^2} \right) e^2 \right.$$

$$+ \frac{1}{4} \left(a' \frac{dC_0}{da'} + \frac{a'^2}{2} \frac{d^2 C_0}{da'^2} \right) e'^2$$

$$+ \frac{1}{4} \left(4 C_1 + 2a \frac{dC_1}{da} + 2a' \frac{dC_1}{da'} + aa' \frac{d^2 C_1}{da \, da'} \right) ee' \cos (\varpi - \varpi')$$

$$- \frac{1}{8} \left(a^2 D_0 + a \frac{dC_0}{da} \right) \tan^2 i - \frac{1}{8} \left(a'^2 D_0 + a' \frac{dC_0}{da'} \right) \tan^2 i''$$

$$\left. + \frac{1}{4} aa' D_1 \tan i \tan i'' \cos (\Omega - \Omega') + \ldots \right\} .$$

We shall hereafter be able to simplify this expression.

48. We have seen that the remaining terms of R are of the form $P \cos \{(pn \pm qn') t + Q\}$: if then values of p and

q could be found such that $pn \pm qn' = 0$, this term, being independent of t explicitly, would form an additional term in F. No instance of this, however, occurs among the planets.

49. In consequence of the extreme smallness of the excentricities and inclinations of the planetary orbits, terms in R of orders higher than the second may in general be neglected: but it sometimes happens, as in the Lunar Theory, that higher terms become sensible through the process of integration. This we shall consider in a subsequent chapter, but the following proposition has an important bearing on the subject.

50. *The principal part of the coefficient of a term in* R *of the form* $P \cos \{(pn - qn')t + Q\}$ *is of the order* p ~ q.

DEF. By the principal part of the coefficient is meant that part of P which is of lowest dimensions in e, e', $\tan i$, $\tan i'$.

If we return to the expression for R in Art. 44, it will be seen that in order to obtain the general term it will be necessary to multiply the product of the general terms of the expansions for u^a, u'^β, v^γ, v'^δ, z^e, z'^ζ by $\cos k\phi$ or $\sin k\phi$.

Now (1) in the expansions of u, u', v, v', z, z' the following law is observed to hold:—The number which multiplies $nt + \epsilon$ or $n't + \epsilon'$ in the argument of any term represents the order of the principal part of the coefficient of that term.

(2) The same holds good in any power of u, u', v, v', z, or z'. For, consider a term $P \cos (pnt + q)$ in u^s. It can only have arisen in the following ways; partly from the multiplication of two terms in u of which the arguments are $lnt + \lambda$ and $mnt + \mu$, where $l + m = p$; and partly from such as have the arguments $l'nt + \lambda'$ and $m'nt + \mu'$, where $l' \sim m' = p$. In the former case the order of the coefficient will be $l + m$, which equals p, in the latter it will be $l' + m'$, and this is

greater than p. Hence the principal part of the coefficient of a term $P \cos (pnt + q)$ in u^2, will be of the order p.

Since then the law holds in u^2, it may be shewn in like manner to hold in the product of u^2 and u, i.e. in u^3. Thus it may be proved for any power of u. In like manner it may be shewn to hold for any powers of u', v, v', z, or z'.

(3) The same law is true for the product of any powers of u, v, z; and likewise for the product of any powers of u', v', z'. This may be proved by a method similar to that of (2).

(4) In the product of any powers of u, u', v, v', z and z', the order of the principal part of the coefficient is the arithmetical sum of the multipliers of nt and $n't$.

For let us consider a term $M \cos \{(ln \pm l'n') t + N\}$. Now this must evidently have arisen from the multiplication of $L \cos (lnt + \lambda)$ with $L' \cos (l'n't + \lambda')$, or of $L \sin (lnt + \lambda)$ with $L' \sin (l'n't + \lambda')$, where by (3) L is of the order l and L' of the order l'. Hence M will be of the order $l + l'$.

Now any term in the development of R of the form $P \cos \{(pn - qn') t + Q\}$ must have arisen partly from the multiplication of $P_1 \genfrac{}{}{0pt}{}{\cos}{\sin} k\phi$, or as it may be written

$$P_1 \genfrac{}{}{0pt}{}{\cos}{\sin} \{(kn - kn') t + Q_1\}$$

with $\qquad P_2 \genfrac{}{}{0pt}{}{\cos}{\sin} \{[(p-k)n - (q-k)n'] t + Q_2\}$,

and partly from its multiplication with

$$P_3 \genfrac{}{}{0pt}{}{\cos}{\sin} \{[(p+k)n - (q+k)n'] t + Q_3\},$$

where k is any positive integer or zero, P_1 is a function of a and a' only, and P_2, P_3 are functions of the excentricities and

DEVELOPMENT OF THE DISTURBING FUNCTION. 47

inclinations, such that the orders of their principal parts are given by law (4). Hence the order of the principal part of P will be equal to the lesser of those of P_2 and P_3.

Now the order of the principal part of P_2 will be the least value of which the arithmetical sum of $p \sim k$ and $q \sim k$ is susceptible, for different values of k.

(i) Suppose k intermediate to p and q; then this sum
$$= p \sim k + q \sim k = p \sim q;$$

(ii) Suppose k not greater than the smaller of p and q; then this sum $= p + q - 2k$, the least value of which (by putting k equal to the smaller of p and q) $= p \sim q$;

(iii) Suppose k not less than the greater of p and q; then this sum $= 2k - p - q$, the least value of which (by putting k equal to the greater of p and q) $= p \sim q$.

Thus $p \sim q$ is the order of the principal part of P_2. That of P_3 will be the least value of which $p + k + q + k$ is susceptible, i.e. $p + q$.

Hence it appears that the order of the principal part of P is $p \sim q$.

51. *The principal part of the coefficient of a term in* R, *of the form* P $\cos \{(pn + qn') t + Q\}$ *is of the order* p + q.

This term arises from the multiplication of such terms as

$$P_1 {\cos \atop \sin} \{(kn - kn') t + Q_1\},$$

with $\quad P_2 {\cos \atop \sin} \{[(p-k) n + (q+k) n'] t + Q_2\},$

and $\quad P_3 {\cos \atop \sin} \{[(p+k) n + (q-k) n'] t + Q_3\},$

and as in the last Article, the order of the principal part of P will be equal to the lesser of those of P_2 and P_3.

48   PLANETARY THEORY.

Now the order of the principal part of P_2 will be the least value which the arithmetical sum of $p-k$ and $q+k$ can assume, for different values of k.

(i) Suppose k less than p; then this sum
$$= p - k + q + k = p + q.$$

(ii) Suppose k not less than p; then this sum
$$= k - p + q + k,$$
the least value of which (by putting k equal to p) $= p + q$.

Similarly it may be shewn that $p+q$ will be the order of the principal part of P_3.

Hence it follows that $p+q$ will be the order of the principal part of P.

In Art. 44 we have assumed that $(a^2 + a'^2 - 2aa' \cos \phi)^{-s}$ can be expanded in a series of cosines of ϕ and its multiples, we shall now give a proof of this and shew how the coefficients may be calculated.

52. *To shew that* $(a^2 + a'^2 - 2aa' \cos \phi)^{-s}$ *can be expanded in a series of cosines of multiples of* ϕ.

Suppose a greater than a', and for $\dfrac{a'}{a}$ write α; then

$$(a^2 + a'^2 - 2aa' \cos \phi)^{-s} = a^{-2s}(1 + \alpha^2 - 2\alpha \cos \phi)^{-s}$$

$$= a^{-2s}\{1 + \alpha^2 - \alpha(e^{\phi\sqrt{-1}} + e^{-\phi\sqrt{-1}})\}^{-s}$$

$$= a^{-2s}(1 - \alpha e^{\phi\sqrt{-1}})^{-s}(1 - \alpha e^{-\phi\sqrt{-1}})^{-s}$$

$$= a^{-2s}\left\{1 + s\alpha e^{\phi\sqrt{-1}} + \frac{s(s+1)}{\lfloor 2} \alpha^2 e^{2\phi\sqrt{-1}}\right.$$

$$\left. + \frac{s(s+1)(s+2)}{\lfloor 3} \alpha^3 e^{3\phi\sqrt{-1}} + \ldots\right\}$$

DEVELOPMENT OF THE DISTURBING FUNCTION. 49

$$\times \left\{ 1 + sae^{-\phi\sqrt{-1}} + \frac{s(s+1)}{\underline{|2}} a^2 e^{-2\phi\sqrt{-1}} \right.$$

$$\left. + \frac{s(s+1)(s+2)}{\underline{|3}} a^3 e^{-3\phi\sqrt{-1}} + \ldots \right\}$$

$$= a^{-2s} \left[1 + s^2 a^2 + \left\{ \frac{s(s+1)}{\underline{|2}} \right\}^2 a^4 \right.$$

$$\left. + \left\{ \frac{s(s+1)(s+2)}{\underline{|3}} \right\}^2 a^6 + \ldots \right]$$

$$+ 2a^{-2s} \left\{ sa + s \cdot \frac{s(s+1)}{\underline{|2}} a^3 \right.$$

$$\left. + \frac{s(s+1)}{\underline{|2}} \cdot \frac{s(s+1)(s+2)}{\underline{|3}} a^5 + \ldots \right\} \left(\frac{e^{\phi\sqrt{-1}} + e^{-\phi\sqrt{-1}}}{2} \right)$$

$$+ \ldots\ldots\ldots,$$

where the coefficients of $e^{k\phi\sqrt{-1}}$ and $e^{-k\phi\sqrt{-1}}$ will always be equal. Hence we may write

$$(a^2 + a'^2 - 2aa' \cos \phi)^{-s}$$

$$= \frac{1}{2} A_0 + A_1 \cos \phi + A_2 \cos 2\phi + \ldots + A_k \cos k\phi + \ldots,$$

where A_0, A_1, &c., are functions of a and a'. The series which they represent will be always convergent provided α is less than unity, or a greater than a'. If a be less than a', we have only to interchange a and a' in the above, so that α will then denote the ratio of a to a'.

53. *To calculate* C_0 *and* C_1.

In the preceding article, let $s = \frac{1}{2}$; then

$$C_0 = 2a^{-1} \left\{ 1 + \left(\frac{1}{2}\right)^2 a^2 + \left(\frac{1 \cdot 3}{2 \cdot 4}\right)^2 a^4 + \left(\frac{1 \cdot 3 \cdot 5}{2 \cdot 4 \cdot 6}\right)^2 a^6 + \ldots \right\},$$

$$C_1 = 2a^{-1} \left\{ \frac{1}{2} \alpha + \frac{1}{2} \cdot \frac{1 \cdot 3}{2 \cdot 4} a^3 + \frac{1 \cdot 3}{2 \cdot 4} \cdot \frac{1 \cdot 3 \cdot 5}{2 \cdot 4 \cdot 6} a^5 + \ldots \right\}.$$

C.P.T. 4

Unless a be small, these series will converge too slowly to be practically useful. More convergent series might be obtained, but according to Pontécoulant (*Système du Monde*, Tome III. p. 81), it is more convenient to employ elliptic integrals for the purpose, in the manner we proceed to explain. We have

$$\frac{1}{(a^2 + a'^2 - 2aa' \cos \phi)^{\frac{1}{2}}} = \frac{1}{2} C_0 + C_1 \cos \phi + C_2 \cos 2\phi + \ldots$$

$$\frac{\cos \phi}{(a^2 + a'^2 - 2aa' \cos \phi)^{\frac{1}{2}}} = \frac{1}{2} C_0 \cos \phi + \frac{1}{2} C_1 (1 + \cos 2\phi)$$

$$+ \frac{1}{2} C_2 (\cos 3\phi + \cos \phi) + \ldots$$

Integrating both sides of these equations with respect to ϕ between the limits 0 and 2π, we obtain

$$\pi C_0 = \int_0^{2\pi} \frac{d\phi}{(a^2 + a'^2 - 2aa' \cos \phi)^{\frac{1}{2}}} = \frac{1}{a} \int_0^{2\pi} \frac{d\phi}{(1 + \alpha^2 - 2\alpha \cos \phi)^{\frac{1}{2}}},$$

$$\pi C_1 = \int_0^{2\pi} \frac{\cos \phi \, d\phi}{(a^2 + a'^2 - 2aa' \cos \phi)^{\frac{1}{2}}} = \frac{1}{a} \int_0^{2\pi} \frac{\cos \phi \, d\phi}{(1 + \alpha^2 - 2\alpha \cos \phi)^{\frac{1}{2}}}.$$

These integrals may be reduced to the standard forms of elliptic functions by assuming

$$\sin (\theta - \phi) = \alpha \sin \theta \quad \ldots\ldots\ldots\ldots\ldots\ldots (1),$$

whence

$$\tan \theta = \frac{\sin \phi}{\cos \phi - \alpha} \quad \ldots\ldots\ldots\ldots\ldots\ldots (2).$$

From (1) $\cos (\theta - \phi) \left(1 - \dfrac{d\phi}{d\theta}\right) = \alpha \cos \theta$;

therefore $\dfrac{d\phi}{d\theta} = \dfrac{\cos (\theta - \phi) - \alpha \cos \theta}{\cos (\theta - \phi)}.$

DEVELOPMENT OF THE DISTURBING FUNCTION.

Now

$$\cos(\theta - \phi) - \alpha \cos\theta = \cos\theta (\cos\phi - \alpha) + \sin\theta \sin\phi$$

$$= \left(\frac{\cos^2\theta}{\sin\theta} + \sin\theta\right)\sin\phi, \text{ by (2)},$$

$$= \frac{\sin\phi}{\sin\theta}$$

$$= \sqrt{\{(\cos\phi - \alpha)^2 + \sin^2\phi\}}, \text{ by (2)},$$

$$= \sqrt{(1 + \alpha^2 - 2\alpha\cos\phi)} \quad \ldots\ldots\ldots (3).$$

Also $\quad \cos(\theta - \phi) = \sqrt{(1 - \alpha^2 \sin^2\theta)} \ldots\ldots\ldots\ldots (4).$

Hence $\quad \dfrac{d\phi}{d\theta} = \sqrt{\left(\dfrac{1 + \alpha^2 - 2\alpha\cos\phi}{1 - \alpha^2 \sin^2\theta}\right)}.$

Again, from equations (3) and (4)

$$\sqrt{(1 + \alpha^2 - 2\alpha\cos\phi)} = \sqrt{(1 - \alpha^2 \sin^2\theta)} - \alpha\cos\theta\ ;$$

therefore $\quad 1 + \alpha^2 - 2\alpha\cos\phi = 1 - \alpha^2 \sin^2\theta$

$$\qquad\qquad + \alpha^2 \cos^2\theta - 2\alpha\cos\theta \sqrt{(1 - \alpha^2 \sin^2\theta)},$$

$$2\alpha\cos\phi = 2\alpha^2 \sin^2\theta + 2\alpha\cos\theta \sqrt{(1 - \alpha^2 \sin^2\theta)},$$

or $\quad \cos\phi = \alpha\sin^2\theta + \cos\theta \sqrt{(1 - \alpha^2 \sin^2\theta)}.$

Now as ϕ increases from 0 up to 2π, θ also increases from 0 to 2π; hence

$$C_0 = \frac{1}{a\pi}\int_0^{2\pi} \frac{d\phi}{\sqrt{(1 + \alpha^2 - 2\alpha\cos\phi)}}$$

$$= \frac{1}{a\pi}\int_0^{2\pi} \frac{d\theta}{\sqrt{(1 - \alpha^2 \sin^2\theta)}},$$

$$C_1 = \frac{1}{a\pi}\int_0^{2\pi} \frac{\cos\phi\, d\phi}{\sqrt{(1 + \alpha^2 - 2\alpha\cos\phi)}}$$

4—2

$$= \frac{1}{a\pi} \int_0^{2\pi} \frac{a \sin^2\theta \, d\theta}{\sqrt{(1-a^2 \sin^2\theta)}} + \frac{1}{a\pi} \int_0^{2\pi} \cos\theta \, d\theta$$

$$= \frac{1}{aa\pi} \left\{ \int_0^{2\pi} \frac{d\theta}{\sqrt{(1-a^2 \sin^2\theta)}} - \int_0^{2\pi} \sqrt{(1-a^2 \sin^2\theta)} \, d\theta \right\}.$$

Hence with the usual notation for elliptic integrals (see Todhunter's *Integral Calculus*, Art. 222),

$$C_0 = \frac{1}{a\pi} F(a, 2\pi) = \frac{4}{a\pi} F\left(a, \frac{\pi}{2}\right),$$

$$C_1 = \frac{1}{a'\pi} \{F(a, 2\pi) - E(a, 2\pi)\}$$

$$= \frac{4}{a'\pi} \left\{ F\left(a, \frac{\pi}{2}\right) - E\left(a, \frac{\pi}{2}\right) \right\}.$$

The numerical values of $F\left(a, \dfrac{\pi}{2}\right)$ and $E\left(a, \dfrac{\pi}{2}\right)$ may be found from Legendre's tables of elliptic functions.

54. *Given C_k and C_{k-1} to obtain C_{k+1}.*

We have

$$(a^2 + a'^2 - 2aa' \cos\phi)^{-\frac{1}{2}} = \frac{1}{2} C_0 + C_1 \cos\phi + \ldots + C_k \cos k\phi + \ldots ;$$

differentiating with respect to ϕ,

$$aa' \sin\phi \, (a^2 + a'^2 - 2aa' \cos\phi)^{-\frac{3}{2}} = C_1 \sin\phi + 2C_2 \sin 2\phi + \ldots$$
$$+ k C_k \sin k\phi + \ldots ;$$

therefore $\quad aa' \sin\phi \left(\dfrac{1}{2} C_0 + C_1 \cos\phi + \ldots \right)$

$$= (a^2 + a'^2 - 2aa' \cos\phi)(C_1 \sin\phi + 2C_2 \sin 2\phi + \ldots);$$

equating coefficients of $\sin k\phi$,

$$\frac{1}{2} aa'(C_{k-1} - C_{k+1}) = k(a^2 + a'^2) C_k - aa' \{(k-1) C_{k-1} + (k+1) C_{k+1}\};$$

whence $\quad C_{k+1} = \dfrac{2k}{2k+1} \dfrac{a^2 + a'^2}{aa'} C_k - \dfrac{2k-1}{2k+1} C_{k-1}.$

55. *Given C_k and C_{k+1} to obtain D_k.*

As in the last article, we have

$$aa' \sin\phi\, (a^2 + a'^2 - 2aa' \cos\phi)^{-\frac{3}{2}} = C_1 \sin\phi + 2C_2 \sin 2\phi + \ldots + kC_k \sin k\phi + \ldots ;$$

therefore $aa' \sin\phi \left(\dfrac{1}{2} D_0 + D_1 \cos\phi + D_2 \cos 2\phi + \ldots\right)$

$$= C_1 \sin\phi + 2C_2 \sin 2\phi + \ldots ;$$

equating coefficients of $\sin k\phi$,

$$2k C_k = aa' (D_{k-1} - D_{k+1}) \quad\ldots\ldots\ldots\ldots\ldots (1),$$

writing $k+1$ for k,

$$2(k+1) C_{k+1} = aa' (D_k - D_{k+2}) \quad\ldots\ldots\ldots\ldots (2).$$

Again,

$$(a^2 + a'^2 - 2aa' \cos\phi)^{-\frac{3}{2}} = \dfrac{1}{2} D_0 + D_1 \cos\phi + D_2 \cos 2\phi + \ldots,$$

and

$$(a^2 + a'^2 - 2aa' \cos\phi)^{-\frac{1}{2}} = \dfrac{1}{2} C_0 + C_1 \cos\phi + C_2 \cos 2\phi + \ldots ;$$

therefore $\quad \dfrac{1}{2} C_0 + C_1 \cos\phi + \ldots$

$$= (a^2 + a'^2 - 2aa' \cos\phi) \left(\dfrac{1}{2} D_0 + D_1 \cos\phi + \ldots\right):$$

equating coefficients of $\cos k\phi$,

$$C_k = (a^2 + a'^2) D_k - aa' (D_{k-1} + D_{k+1}) \quad\ldots\ldots\ldots (3),$$

writing $k+1$ for k,

$$C_{k+1} = (a^2 + a'^2) D_{k+1} - aa' (D_k + D_{k+2}) \quad\ldots\ldots\ldots (4).$$

Eliminating D_{k-1} between (1) and (3),

$$(2k+1) C_k = (a^2 + a'^2) D_k - 2aa' D_{k+1} \quad\ldots\ldots\ldots (5).$$

Eliminating D_{k+2} between (2) and (4),

$$(2k+1)\, C_{k+1} = -(a^2 + a'^2)\, D_{k+1} + 2aa' D_k \dots \dots (6).$$

Finally, eliminating D_{k+1} between (5) and (6),

$$(2k+1)\{(a^2+a'^2)\, C_k - 2aa' C_{k+1}\} = \{(a^2+a'^2)^2 - 4a^2 a'^2\}\, D_k,$$

or $$D_k = \frac{2k+1}{(a^2-a'^2)^2}\{(a^2+a'^2)\, C_k - 2aa' C_{k+1}\}.$$

56. *To calculate the successive differential coefficients of* C_k *and* D_k *with respect to* a *and* a'.

We have

$$(a^2 + a'^2 - 2aa'\cos\phi)^{-\frac{1}{2}} = \frac{1}{2}C_0 + C_1 \cos\phi + C_2 \cos 2\phi + \dots$$
$$+ C_k \cos k\phi + \dots :$$

differentiating with respect to a,

$$-(a - a'\cos\phi)(a^2 + a'^2 - 2aa'\cos\phi)^{-\frac{3}{2}} = \frac{1}{2}\frac{dC_0}{da} + \frac{dC_1}{da}\cos\phi + \dots$$
$$+ \frac{dC_k}{da}\cos k\phi + \dots;$$

substituting for $(a^2 + a'^2 - 2aa'\cos\phi)^{-\frac{3}{2}}$ its expression in series,

$$-(a - a'\cos\phi)\left(\frac{1}{2}D_0 + D_1 \cos\phi + \dots + D_k \cos k\phi + \dots\right)$$
$$= \frac{1}{2}\frac{dC_0}{da} + \frac{dC_1}{da}\cos\phi + \dots + \frac{dC_k}{da}\cos k\phi + \dots;$$

equating coefficients of $\cos k\phi$,

$$\frac{dC_k}{da} = -aD_k + \frac{a'}{2}(D_{k-1} + D_{k+1}).$$

By giving to k in succession the values 1, 2, 3, &c., those of $\dfrac{dC_1}{da}$, $\dfrac{dC_2}{da}$, &c. may be found, the right-hand member

DEVELOPMENT OF THE DISTURBING FUNCTION. 55

being calculated by the formula of the last article. By equating the parts independent of ϕ, we obtain

$$\frac{dC_0}{da} = -aD_0 + a'D_1.$$

The value of $\frac{dD_k}{da}$ may be found by differentiating the expression for D_k in Art. 55, and substituting for $\frac{dC_k}{da}$ and $\frac{dC_{k+1}}{da}$ their values as given by the present article.

The successive differential coefficients of C_k and D_k with respect to a may be obtained from the expressions for $\frac{dC_k}{da}$ and $\frac{dD_k}{da}$ by simple differentiation and substitution.

57. We might determine in the same way the successive differential coefficients of C_k and D_k with respect to a'; but when those with respect to a have been found, the former may be derived from them, as we proceed to shew. On examining the expansion of $(a^2 + a'^2 - 2aa'\cos\phi)^{-s}$ in Art. 52, it will be seen that A_k is a homogeneous function of a and a of $-2s$ dimensions. Hence C_k and D_k are homogeneous functions of a and a', the former of -1, the latter of -3 dimensions. It follows that $\frac{dC_k}{da}$, $\frac{dD_k}{da}$ will be homogeneous functions of -2 and -4 dimensions respectively; and so on. Now by a known property of such functions

$$a\frac{dC_k}{da} + a'\frac{dC_k}{da'} = -C_k,$$

which determines $\frac{dC_k}{da'}$:

$$a\frac{d^2C_k}{da^2} + a'\frac{d^2C_k}{da\,da'} = -2\frac{dC_k}{da},$$

which determines $\dfrac{d^2C_k}{da\,da'}$:

$$a'\frac{d^2C_k}{da'^2} + a\frac{d^2C_k}{da\,da'} = -2\frac{dC_k}{da'},$$

which determines $\dfrac{d^2C_k}{da'^2}$: and thus all the differential coefficients of C_k may be determined.

In like manner all the successive differential coefficients of D_k may be calculated.

We are now in a position to simplify the expression for F. We have (Art. 47.)

$$F = m'\left\{\frac{C_0}{2} + \frac{1}{4}\left(a\frac{dC_0}{da} + \frac{a^2}{2}\frac{d^2C_0}{da^2}\right)e^2\right.$$

$$+ \frac{1}{4}\left(a'\frac{dC_0}{da'} + a'^2\frac{d^2C_0}{da'^2}\right)e'^2$$

$$+ \frac{1}{4}\left(4C_1 + 2a\frac{dC_1}{da} + 2a'\frac{dC_1}{da'} + aa'\frac{d^2C_1}{da\,da'}\right)ee'\cos(\varpi - \varpi')$$

$$- \frac{1}{8}\left(a^2 D_0 + a\frac{dC_0}{da}\right)\tan^2 i - \frac{1}{8}\left(a'^2 D_0 + a'\frac{dC_0}{da'}\right)\tan^2 i''$$

$$\left. + \frac{1}{4}aa'D_1 \tan i \tan i'' \cos(\Omega - \Omega') + \ldots\right\}.$$

The following proposition will be found useful.

58. To shew that $\dfrac{d^2C_0}{da\,da'} = -D_1$, and that $\dfrac{d^2C_1}{da\,da'} = -D_0$.

We have

$$\frac{1}{2}C_0 + C_1\cos\phi + C_2\cos 2\phi + \ldots = (a^2 + a'^2 - 2aa'\cos\phi)^{-\frac{1}{2}};$$

therefore

$$\frac{1}{2}\frac{dC_0}{da} + \frac{dC_1}{da}\cos\phi + \frac{dC_2}{da}\cos 2\phi + \ldots$$

$$= -(a - a'\cos\phi)(a^2 + a'^2 - 2aa'\cos\phi)^{-\frac{3}{2}};$$

DEVELOPMENT OF THE DISTURBING FUNCTION.

therefore

$$\frac{1}{2}\frac{d^2C_0}{da\,da'} + \frac{d^2C_1}{da\,da'}\cos\phi + \ldots = \cos\phi\,(a^2 + a'^2 - 2aa'\cos\phi)^{-\frac{3}{2}}$$
$$+ 3\,(a - a'\cos\phi)(a' - a\cos\phi)(a^2 + a'^2 - 2aa'\cos\phi)^{-\frac{5}{2}}$$
$$= \cos\phi\,(a^2 + a'^2 - 2aa'\cos\phi)^{-\frac{3}{2}}$$
$$+ 3\,\{aa'(1 + \cos^2\phi) - (a^2 + a'^2)\cos\phi\}(a^2 + a'^2 - 2aa'\cos\phi)^{-\frac{5}{2}}$$
$$= \cos\phi\,(a^2 + a'^2 - 2aa'\cos\phi)^{-\frac{3}{2}}$$
$$+ 3\,\{aa'\sin^2\phi - \cos\phi\,(a^2 + a'^2 - 2aa'\cos\phi)\}$$
$$(a^2 + a'^2 - 2aa'\cos\phi)^{-\frac{5}{2}}$$
$$= -2\cos\phi\,(a^2 + a'^2 - 2aa'\cos\phi)^{-\frac{3}{2}}$$
$$+ 3aa'\sin^2\phi\,(a^2 + a'^2 - 2aa'\cos\phi)^{-\frac{5}{2}}.$$

Now [1] $(a^2 + a'^2 - 2aa'\cos\phi)^{-\frac{3}{2}} = \frac{1}{2}D_0 + D_1\cos\phi + \ldots;$

differentiating with respect to ϕ

$$3aa'\sin\phi\,(a^2 + a'^2 - 2aa'\cos\phi)^{-\frac{5}{2}}$$
$$= D_1\sin\phi + 2D_2\sin 2\phi + \ldots;$$

therefore $\dfrac{1}{2}\dfrac{d^2C_0}{da\,da'} + \dfrac{d^2C_1}{da\,da'}\cos\phi + \ldots$

$$= -2\cos\phi\left(\frac{1}{2}D_0 + D_1\cos\phi + \ldots\right)$$
$$+ \sin\phi\,(D_1\sin\phi + 2D_2\sin 2\phi + \ldots),$$

whence, equating the parts independent of ϕ, and also the coefficients of $\cos\phi$

$$\frac{1}{2}\frac{d^2C_0}{da\,da'} = -D_1 + \frac{D_1}{2} = -\frac{D_1}{2},$$

or $\qquad\qquad\dfrac{d^2C_0}{da\,da'} = -D_1;$

and $\qquad\qquad\dfrac{d^2C_1}{da\,da'} = -D_0.$

59. Since $\dfrac{dC_0}{da}$ is a homogeneous function of a and a' of -2 dimensions,

$$a \frac{d^2C_0}{da^2} + a' \frac{d^2C_0}{da\,da'} = -2 \frac{dC_0}{da};$$

therefore

$$a \frac{dC_0}{da} + \frac{a^2}{2} \frac{d^2C_0}{da^2} = -\frac{aa'}{2} \frac{d^2C_0}{da\,da'}$$

$$= \frac{1}{2} aa' D_1.$$

Similarly, $\quad a' \dfrac{dC_0}{da'} + \dfrac{a'^2}{2} \dfrac{d^2C_0}{da'^2} = \dfrac{1}{2} aa' D_1.$

Hence the coefficients of e^2 and e'^2 in the expression for F are each equal to $\dfrac{1}{8} aa' D_1$.

Again, since C_1 is a homogeneous function of a and a' of -1 dimensions,

$$a \frac{dC_1}{da} + a' \frac{dC_1}{da'} = -C_1;$$

hence the coefficient of $ee' \cos(\varpi - \varpi')$

$$= \frac{1}{4} (2C_1 - aa' D_0);$$

but (Art. 55) $\quad 2k C_k = aa' (D_{k-1} - D_{k+1});$

therefore, making $\quad k = 1,$

$$2C_1 = aa' (D_0 - D_2);$$

hence the coefficient of $ee' \cos(\varpi - \varpi')$

$$= -\frac{1}{4} aa' D_2.$$

Again, (Art. 56)

$$\frac{dC_0}{da} = -aD_0 + a^2 D_1;$$

therefore $$a^2 D_0 + a \frac{dC_0}{da} = aa' D_1.$$

Similarly, $$a'^2 D_0 + a' \frac{dC_0}{da'} = aa' D_1.$$

Hence the coefficients of $\tan^2 i$ and $\tan^2 i''$ are each equal to $-\frac{1}{8} aa' D_1$.

Finally, the expression for F becomes

$$F = m' \left\{ \frac{C_0}{2} + \frac{1}{8} aa' D_1 (e^2 + e'^2) - \frac{1}{4} aa' D_2 ee' \cos(\varpi - \varpi') \right.$$
$$\left. - \frac{1}{8} aa' D_1 (\tan^2 i + \tan^2 i'') + \frac{1}{4} aa' D_1 \tan i \tan i'' \cos(\Omega - \Omega') \right.$$
$$\left. + \ldots \right\}.$$

CHAPTER IV.

SECULAR VARIATIONS OF THE ELEMENTS OF THE ORBIT. STABILITY OF THE PLANETARY SYSTEM.

60. WE have seen in the preceding Chapter, that the disturbing function, when developed, consists of two parts; the one independent of the time explicitly, the other involving it under a periodical form: we shall consider separately the effects of these two parts. In the present Chapter our attention will be directed to the first or non-periodical part of R, which we have denoted by F. The inequalities thus produced in the elements of the orbit are termed *secular*, in consequence of their very slow variation.

61. By differentiating the expression for F in Art. 59, with respect to the elements, we obtain

$$\frac{dF}{d\epsilon} = 0,$$

$$\frac{dF}{d\varpi} = \frac{m'}{4} aa' D_2 ee' \sin(\varpi - \varpi'),$$

$$\frac{dF}{de} = \frac{m'}{4} aa' D_1 e - \frac{m'}{4} aa' D_2 e' \cos(\varpi - \varpi'),$$

$$\frac{dF}{d\Omega} = -\frac{m'}{4} aa' D_1 \tan i \tan i' \sin(\Omega - \Omega'),$$

SECULAR VARIATIONS. 61

$$\frac{dF}{di} = -\frac{m'}{4} aa' D_1 \tan i + \frac{m'}{4} aa' D_1 \tan i'' \cos(\Omega - \Omega'),$$

$$\frac{dF}{da} = \text{an expression similar to } F.$$

Substituting these in the formulæ of Art. 39, and neglecting small quantities of orders higher than the second, we have

$$\frac{da}{dt} = 0,$$

$$\frac{de}{dt} = -\frac{m'na^2a'}{4\mu} D_2 e' \sin(\varpi - \varpi'),$$

$$e\frac{d\varpi}{dt} = \frac{m'na^2a'}{4\mu} \{D_1 e - D_2 e' \cos(\varpi - \varpi')\},$$

$$\frac{di}{dt} = \frac{m'na^2a'}{4\mu} D_1 \tan i'' \sin(\Omega - \Omega'),$$

$$\tan i \frac{d\Omega}{dt} = -\frac{m'na^2a'}{4\mu} D_1 \{\tan i - \tan i'' \cos(\Omega - \Omega')\}$$

$$\frac{d\epsilon}{dt} = A + A_1 (e^2 - \tan^2 i) + A_2 (e'^2 - \tan^2 i'')$$

$$+ A_3 ee' \cos(\varpi - \varpi') + A_4 \tan i \tan i'' \cos(\Omega - \Omega'),$$

where in the last expression, A, A_1, &c., have been written to denote certain functions of a and a'.

62. *To calculate approximately the secular variations of the elements of a planet's orbit, in a given time.*

Let a_0, e_0, ϖ_0, &c., be the values of the elements at some given epoch; $a_0 + \delta a^*$, $e_0 + \delta e$, $\varpi_0 + \delta \varpi$, &c., their values after an interval t: then δa, δe, $\delta \varpi$, &c., are the required variations. By Maclaurin's Theorem,

---

* It will be shewn in Art. 64 that δa is always zero.

$$\delta e = \left(\frac{de}{dt}\right)_0 t + \left(\frac{d^2e}{dt^2}\right)_0 \frac{t^2}{\underline{|2}} + \ldots$$

$$\delta \varpi = \left(\frac{d\varpi}{dt}\right)_0 t + \left(\frac{d^2\varpi}{dt^2}\right)_0 \frac{t^2}{\underline{|2}} + \ldots$$

$$\ldots = \ldots$$

which may be carried to any required degree of accuracy, but in practice the first two terms will generally be sufficient.

We have supposed the variations of the elements required at a time t *after* the epoch; if they be required at a time t *before* the epoch, we have only to change the sign of t in the above.

We may remark that $\left(\frac{de}{dt}\right)_0$, $\left(\frac{d\varpi}{dt}\right)_0$, &c. are of the order of the disturbing force, since they involve the first power of m'; $\left(\frac{d^2e}{dt^2}\right)_0$, $\left(\frac{d^2\varpi}{dt^2}\right)_0$, &c. will involve m'^2 and be of the second order; and so on.

In the short period of one year all terms after the first may be neglected, so that putting $t = 1$, we have

$$\delta e = \left(\frac{de}{dt}\right)_0, \text{ &c.}$$

Hence the coefficient of t in the above formulæ is called the *annual váriation*.

63. Since the elements of the planetary orbits are continually changing, it will be interesting to shew that the dimensions of these orbits, and their inclinations to the ecliptic, nevertheless fluctuate between very narrow limits. This constitutes what is termed the Stability of the Planetary System: in order to establish it, it will be necessary to prove the stability (i) of the mean distances, (ii) of the excentricities, (iii) of the inclinations.

SECULAR VARIATIONS.

64. *To prove the stability of the mean distance of the planets from the Sun, and of their mean motions.*

By Art. 61 $\frac{da}{dt} = 0$, so that a is constant. . Now it will be shewn in a subsequent chapter, (see Art. 91), that to the first order of the disturbing force, the periodical terms of R can produce only periodical variations*; consequently to this order, the mean distance is susceptible of no permanent change. The same is true of the mean motion n, since it $= \frac{\mu^{\frac{1}{2}}}{a^{\frac{3}{2}}}$, and μ does not alter. We are hereby assured of the impossibility of any of the bodies of our system ever leaving it, in consequence of the disturbances it may experience from the other bodies; and this secures the general permanence of the whole, by keeping the mean distances and periodic times perpetually fluctuating between certain limits (very restricted ones) which they can never exceed or fall short of.

This result may easily be extended to all orders of the excentricities and inclinations: for since $nt + \epsilon$ always occurs in R as one symbol, ϵ cannot occur in F because t does not, so that $\frac{dF}{d\epsilon}$, and therefore $\frac{da}{dt}$ is zero.

65. *To prove the stability of the excentricities of the planetary orbits.*

We will first consider the case of two planets only. By Art. 61,

$$\frac{de}{dt} = -\frac{m'na^2a'}{4\mu} D_2 e' \sin(\varpi - \varpi').$$

---

* This result is also true when the square of the disturbing force is included: for the demonstration the reader is referred to Pontécoulant's *Système du Monde*, or to Laplace's *Mécanique Céleste*.

Similarly, $\dfrac{de'}{dt} = -\dfrac{mn'a'^2 a}{4\mu} D_2' e \sin(\varpi' - \varpi)$.

Now since D_2 is the coefficient of $\cos 2\phi$ in the development of
$$(a^2 + a'^2 - 2aa' \cos \phi)^{-\frac{1}{2}},$$
an expression in which a and a' are similarly involved, it follows that
$$D_2' = D_2.$$

Hence, multiplying the above equations by $\dfrac{m}{na} e$, $\dfrac{m'}{n'a'} e'$, respectively, and adding, we have
$$\frac{m}{na} e \frac{de}{dt} + \frac{m'}{n'a'} e' \frac{de'}{dt} = 0;$$
therefore, since a experiences no secular variation,
$$\frac{m}{na} e^2 + \frac{m'}{n'a'} e'^2 = C.$$

A similar equation holds for any number of planets. Replacing for convenience $\dfrac{aa' D_2}{4\mu}$ by (a, a'), we have

$$\frac{de}{dt} = -m'na (a, a') e' \sin(\varpi - \varpi')$$
$$\qquad - m''na (a, a'') e'' \sin(\varpi - \varpi'') - \ldots$$
$$\frac{de'}{dt} = -mn'a' (a', a) e \sin(\varpi' - \varpi)$$
$$\qquad - m''n'a' (a', a'') e'' \sin(\varpi' - \varpi'') - \ldots$$
$$\frac{de''}{dt} = -mn''a'' (a'', a) e \sin(\varpi'' - \varpi)$$
$$\qquad - m'n''a'' (a'', a') e' \sin(\varpi'' - \varpi') - \ldots$$
$$\ldots = \ldots\ldots\ldots\ldots$$

Since $D_2' = D_2$, it follows that $(a, a') = (a', a)$: hence multiplying these equations by $\dfrac{m}{na} e$, $\dfrac{m'}{n'a'} e'$, &c., and adding, we obtain

SECULAR VARIATIONS.

$$\frac{m}{na}e\frac{de}{dt} + \frac{m'}{n'a'}e'\frac{de'}{dt} + \frac{m''}{n''a''}e''\frac{de''}{dt} + \ldots = 0;$$

whence by integration

$$\Sigma\left(\frac{m}{na}e^2\right) = C.$$

Now observation shews that all the planets revolve round the sun in the same direction, so that the mean motions n, n', n'', &c. are of uniform sign. Hence all the terms of the left-hand member of the above equation are positive.

We learn also from observation that the excentricities of the planetary orbits are at present very small indeed, with the exception of the Asteroids, the masses of which are very small. Hence the constant must be small. Since, then, all the terms of the first side of the equation are positive, and their sum always equals a small constant; it follows that every term is small, and therefore that the excentricities are always small*.

66. *To prove the stability of the inclinations of the planes of the planetary orbits.*

By Art. 61 $\dfrac{di}{dt} = \dfrac{m'na^2a'}{4\mu} D_{,} \tan i' \sin(\Omega - \Omega')$.

Similarly, $\dfrac{di''}{dt} = \dfrac{mn'a'^2a}{4\mu} D_{,}' \tan i \sin(\Omega' - \Omega)$.

As in the last article, it may be shewn that $D_{,}' = D_{,}$. Hence, multiplying the above equations by

$$\frac{m}{na}\tan i, \quad \frac{m'}{n'a'}\tan i',$$

---

* It should be noticed that the above is satisfactory only for those planets whose masses are considerable, which is the case with Jupiter and Saturn; but the stability of the excentricities is not confined to these planets. For a complete discussion of the subject the reader is referred to Pontécoulant's *Système du Monde*, or to Laplace's *Mécanique Céleste*.

respectively, and adding, we have
$$\frac{m}{na} \tan i \cdot \frac{di}{dt} + \frac{m'}{n'a'} \tan i'' \frac{di''}{dt} = 0,$$
or to the same order of approximation,
$$\frac{m}{na} \tan i \cdot \frac{d(\tan i)}{dt} + \frac{m'}{n'a'} \tan i'' \frac{d(\tan i'')}{dt} = 0;$$
therefore
$$\frac{m}{na} \tan^2 i + \frac{m'}{n'a'} \tan^2 i'' = C.$$

A similar equation would (as in the case of the excentricities) be true for any number of planets. Now the inclinations of the planetary orbits to the ecliptic are at present very small: hence if we take for our fixed plane of reference a plane coinciding with the present position of the ecliptic, it follows, as in Art. 65, that their inclinations to this plane must always remain very small*.

67. The stability of the excentricities and inclinations may also be established as follows.

By conservation of areas
$$\Sigma (mh \cos i) = \text{const.},$$
or since
$$h^2 = \mu a (1 - e^2) = (M + m) a (1 - e^2),$$
if M denote the mass of the sun, we have
$$\Sigma \left\{ m\sqrt{(Ma)} \left(1 + \frac{1}{2}\frac{m}{M} + \ldots\right) \left(1 - \frac{e^2}{2} + \ldots\right) \left(1 - \frac{i^2}{2} + \ldots\right) \right\} = \text{const.}$$

Since a is constant, if we neglect $\frac{m^2}{\sqrt{M}}$, and the fourth powers of the excentricities and inclinations, this may be written
$$\Sigma (m \sqrt{a} e^2) + \Sigma (m \sqrt{a} i^2) = C,$$

---

* We may remark that the above demonstration, like that of the preceding article, is applicable only to the case of planets of considerable mass.

or to the same order of approximation

$$\Sigma (m \sqrt{a} e^2) + \Sigma (m \sqrt{a} \tan^2 i) = C.$$

Since we know from observation that all the planets revolve round the sun in the same direction, all the radicals in this equation must be taken with the same sign. Also, since the excentricities and inclinations are at present very small, the constant must be small. Hence it follows, as in Arts. 65 and 66, that the excentricities and inclinations must always remain very small.

68. It may be observed that the result of the preceding article proves the stability of the excentricities and inclinations as far as the third order of small quantities, while in Arts. 65 and 66 it was only established to the second order. We will now shew that if small quantities of orders higher than the second be neglected, the equation of the preceding article includes those of Arts. 65 and 66.

On referring to Art. 61, it will be seen that to the second order, the excentricities and inclinations are given by equations independent the one of the other. Each must therefore be the same as if the other did not exist. Hence in the equation of the preceding article, if we make successively $i = 0$ and $e = 0$, we have

$$\Sigma (m \sqrt{a} e^2) = C, \quad \Sigma (m \sqrt{a} \tan^2 i) = C;$$

or, since $\sqrt{a} = \dfrac{\sqrt{\mu}}{na}$,

$$\Sigma \left(\frac{m}{na} e^2 \right) = C, \quad \Sigma \left(\frac{m}{na} \tan^2 i \right) = C,$$

which agree with Arts. 65 and 66.

69. From the results of the preceding articles, we draw the following remarkable conclusion: *The fact that the planets revolve about the sun in the same direction, ensures the stability*

of the planetary system. The converse of this would not necessarily be true, as we shall see in Art. 75: the numerical relations of the dimensions and positions of the orbits of the planets, might be such as to ensure stability, although they revolved in opposite directions. But the above is independent of particular numerical relations.

70. The conclusions at which we have arrived with regard to the stability of the planetary system are of especial interest. In consequence of the changes in the elements it might have been supposed that the orbits would ultimately undergo such alterations in their dimensions as to bring the planets into collision or hurry them into boundless space. Or even if no such violent catastrophe occurred, a derangement of the seasons might seriously have interfered with the physical comfort of man*. But analysis shews (and the results are confirmed when the approximation is carried further,) that unless some unknown cause should operate to produce the contrary, the dimensions and position of the orbits will for ages remain nearly the same as they are at present, i. e. nearly circular in form, and but little inclined to each other, thus affording a beautiful illustration of Gen. viii. 22: "While the earth remaineth, seed-time and harvest, and cold and heat, and summer and winter, and day and night shall not cease."

* See Herschel's *Outlines of Astronomy*.

CHAPTER V.

SECULAR VARIATIONS OF THE ELEMENTS CONTINUED. INTEGRATION OF THE DIFFERENTIAL EQUATIONS.

71. In Art. 62 we have given a method of calculating the secular variations sufficiently accurate for the practical purposes of astronomy, but the equations of Art. 61 admit of actual integration. To this we now proceed.

72. *To integrate the equations for the excentricity and longitude of perihelion.*

We have (Art. 61) for the planet m

$$\frac{de}{dt} = -\frac{m'na^2a'}{4\mu} D_2 e' \sin(\varpi - \varpi'),$$

$$e\frac{d\varpi}{dt} = \frac{m'na^2a'}{4\mu} \{D_1 e - D_2 e' \cos(\varpi - \varpi')\} ;$$

with similar equations for the planet m'.

We shall be able to reduce these to a system of linear differential equations if we assume

$$u = e \sin \varpi, \qquad v = e \cos \varpi,$$
$$u' = e' \sin \varpi', \qquad v' = e' \cos \varpi' ;$$

therefore
$$\frac{du}{dt} = e \cos \varpi \frac{d\varpi}{dt} + \sin \varpi \frac{de}{dt}.$$

Substituting the values of $\frac{d\varpi}{dt}$ and $\frac{de}{dt}$, and writing α for $\frac{m'na^2a'}{4\mu}$, we have

$$\frac{du}{dt} = \alpha(D_1 e \cos \varpi - D_2 e' \cos \varpi')$$
$$= \alpha(D_1 v - D_2 v').$$

Similarly, $\quad \frac{dv}{dt} = \alpha(D_2 u' - D_1 u).$

In like manner for the planet m', writing α' for $\frac{mn'a'^2a}{4\mu}$, we have

$$\frac{du'}{dt} = \alpha'(D_1 v' - D_2 v),$$
$$\frac{dv'}{dt} = \alpha'(D_2 u - D_1 u').$$

The forms of these equations suggest the following particular integrals:

$u = M \sin(gt + \gamma), \quad v = M \cos(gt + \gamma),$
$u' = M' \sin(gt + \gamma), \quad v' = M' \cos(gt + \gamma).$

Substituting these in the differential equations, we obtain from either of the first two

$$gM = \alpha(D_1 M - D_2 M'),$$

and from either of the last two

$$gM' = \alpha'(D_1 M' - D_2 M);$$

eliminating the ratio $M : M'$

$$(g - \alpha D_1)(g - \alpha' D_1) = \alpha\alpha' D_2^2,$$
or $\quad g^2 - (\alpha + \alpha') D_1 g + \alpha\alpha'(D_1^2 - D_2^2) = 0:$

and the roots of this equation will be real and unequal, real and equal, or impossible, according as

$$(\alpha + \alpha')^2 D_1^2 - 4\alpha\alpha'(D_1^2 - D_2^2)$$

is positive, zero, or negative. Now

$$(a+a')^2 D_1^2 - 4aa'(D_1^2 - D_2^2) = (a-a')^2 D_1^2 + 4aa' D_2^2,$$

a positive quantity, since n, n' and therefore a, a' are of like sign. Hence the values of g will be real and unequal: denote them by g_1, g_2, and let γ_1, γ_2; M_1, M_2; M_1', M_2'; be the corresponding values of γ, M, M' respectively. Then the complete solution of the differential equations will be

$$u = M_1 \sin(g_1 t + \gamma_1) + M_2 \sin(g_2 t + \gamma_2),$$
$$v = M_1 \cos(g_1 t + \gamma_1) + M_2 \cos(g_2 t + \gamma_2),$$
$$u' = M_1' \sin(g_1 t + \gamma_1) + M_2' \sin(g_2 t + \gamma_2),$$
$$v' = M_1' \cos(g_1 t + \gamma_1) + M_2' \cos(g_2 t + \gamma_2).$$

Of the constants in these equations, four are arbitrary and must be determined from observation. We have

$$e^2 = u^2 + v^2 = M_1^2 + M_2^2 + 2M_1 M_2 \cos\{(g_1 - g_2)t + \gamma_1 - \gamma_2\},$$
$$\tan \varpi = \frac{u}{v} = \frac{M_1 \sin(g_1 t + \gamma_1) + M_2 \sin(g_2 t + \gamma_2)}{M_1 \cos(g_1 t + \gamma_1) + M_2 \cos(g_2 t + \gamma_2)}:$$

with similar equations for e' and ϖ'.

73. Had we considered a system of several planets, we should have obtained by a similar process

$$e^2 = M_1^2 + M_2^2 + M_3^2 + \ldots + 2M_1 M_2 \cos\{(g_1 - g_2)t + \gamma_1 - \gamma_2\}$$
$$+ 2M_1 M_3 \cos\{(g_1 - g_3)t + \gamma_1 - \gamma_3\} + \ldots$$
$$\tan \varpi = \frac{M_1 \sin(g_1 t + \gamma_1) + M_2 \sin(g_2 t + \gamma_2) + M_3 \sin(g_3 t + \gamma_3) + \ldots}{M_1 \cos(g_1 t + \gamma_1) + M_2 \cos(g_2 t + \gamma_2) + M_3 \cos(g_3 t + \gamma_3) + \ldots},$$

with similar equations for each of the other planets.

74. From the form of the expression for e in Art. 72, the stability of the excentricities, in the case of two planets, may be inferred. We have

$$e^2 = M_1^2 + M_2^2 + 2M_1 M_2 \cos\{(g_1 - g_2)t + \gamma_1 - \gamma_2\}.$$

72 PLANETARY THEORY.

Consequently the excentricity fluctuates between the limits $M_1 + M_2$ and $M_1 \sim M_2$, and since we know from observation that M_1 and M_2 are very small*, it follows that the excentricity must always remain very small.

The period of the changes in the excentricities $= \dfrac{2\pi}{g_1 \sim g_2}$, and is the same for each planet. In the case of Jupiter and Saturn this amounts to 70414 years! The greatest and least excentricities which Jupiter's orbit can attain are ·06036 and ·02606, those of Saturn ·08409 and ·01345; the maximum of each excentricity taking place at the time of the minimum of the other. This follows from the equation

$$\frac{m}{na} e^2 + \frac{m'}{n'a'} e'^2 = C,$$

which has been obtained in Art. 65.

75. It appears from the preceding article that the stability of the excentricities is a consequence of the periodical form of the solution of the differential equations, a result which depends upon the fact that g_1 and g_2 are real and unequal. Now we have seen that in order that this may be the case, it is only necessary that

$$(\alpha + \alpha')^2 D_1^2 - 4\alpha\alpha' (D_1^2 - D_2^2)$$

shall be positive, a condition which might be satisfied if the signs of n, n', and therefore of α, α' were different. In this case, then, the stability would still subsist. Let us however consider what would be the effect of equal or impossible roots to the quadratic from which g is found. In the former case a term would be introduced into $u, u', v,$ and v' proportional to

---

* In the case of Jupiter and Saturn, Sir John Herschel finds that
$M_1 = -\cdot 01715,$ $M_2 = \cdot 04321,$ for Jupiter;
$M_1' = \cdot 04877,$ $M_2' = \cdot 03532,$ for Saturn:
the year 1700 being taken as the epoch. See Article, *Physical Astronomy* in the *Encyclopædia Metropolitana*.

SECULAR VARIATIONS.

the time, and in the latter the periodical terms would be replaced by exponentials. Consequently the excentricities would increase indefinitely with the time, and the stability would no longer subsist.

76. We now proceed to examine the expression which has been obtained in Art. 72, for the longitude of perihelion, viz.

$$\tan \varpi = \frac{M_1 \sin(g_1 t + \gamma_1) + M_2 \sin(g_2 t + \gamma_2)}{M_1 \cos(g_1 t + \gamma_1) + M_2 \cos(g_2 t + \gamma_2)};$$

$$\therefore \frac{d\varpi}{dt} = \frac{g_1 M_1^2 + g_2 M_2^2 + (g_1 + g_2) M_1 M_2 \cos\{(g_1 - g_2) t + \gamma_1 - \gamma_2\}}{M_1^2 + M_2^2 + 2 M_1 M_2 \cos\{(g_1 - g_2) t + \gamma_1 - \gamma_2\}}.$$

The maxima and minima values of ϖ, if such exist, will be found by equating $\dfrac{d\varpi}{dt}$ to zero. Thus

$$\cos\{(g_1 - g_2) t + \gamma_1 - \gamma_2\} = -\frac{g_1 M_1^2 + g_2 M_2^2}{(g_1 + g_2) M_1 M_2}.$$

If this (disregarding sign) be not greater than unity, the perihelion will oscillate, the period of a complete oscillation being the same as that of the excentricities, viz. $\dfrac{2\pi}{g_1 \sim g_2}$; but if, as is the case with Jupiter and Saturn, this be greater than unity, the longitude of perihelion has no maximum or minimum, and the perihelion moves constantly in one direction.

Again,

$$\frac{d\varpi}{dt} = \frac{1}{2} \frac{(g_1 - g_2)(M_1^2 - M_2^2)}{M_1^2 + M_2^2 + 2 M_1 M_2 \cos\{(g_1 - g_2) t + \gamma_1 - \gamma_2\}} + \frac{1}{2}(g_1 + g_2)$$

$$= \frac{(g_1 - g_2)(M_1^2 - M_2^2)}{2 e^2} + \frac{1}{2}(g_1 + g_2).$$

Hence when e is a maximum or minimum, $\dfrac{d\varpi}{dt}$ will be either a maximum or minimum, and the apsidal line will be moving most rapidly or most slowly, different cases occurring according to the signs and magnitudes of the quantities involved.

77. *When the apsidal line oscillates, to find the extent and periods of its oscillations.*

We have (Art. 72)

$$\tan \varpi = \frac{M_1 \sin(g_1 t + \gamma_1) + M_2 \sin(g_2 t + \gamma_2)}{M_1 \cos(g_1 t + \gamma_1) + M_2 \cos(g_2 t + \gamma_2)};$$

therefore $\tan(\varpi - g_1 t - \gamma_1) = \dfrac{\tan \varpi - \tan(g_1 t + \gamma_1)}{1 + \tan \varpi \tan(g_1 t + \gamma_1)}$

$$= \frac{-M_2 \sin \psi}{M_1 + M_2 \cos \psi},$$

if $\psi = (g_1 - g_2) t + \gamma_1 - \gamma_2$.

Also by the last article, if τ be the least positive angle whose cosine is $-\dfrac{g_1 M_1^2 + g_2 M_2^2}{(g_1 + g_2) M_1 M_2}$,

$$e^2 \frac{d\varpi}{dt} = (g_1 + g_2) M_1 M_2 (\cos \psi - \cos \tau).$$

Different cases will occur according to the signs of M_1, M_2, &c. Suppose M_1, M_2 to have different signs, g_1 and g_2 positive, and g_1 greater than g_2. Then ψ increases as t increases, and $\dfrac{d\varpi}{dt}$ will be negative, or the apsidal line will regrede, while $\cos \psi - \cos \tau$ is positive, i.e. so long as ψ is between $2n\pi - \tau$ and $2n\pi + \tau$: $\dfrac{d\varpi}{dt}$ will be positive, or the apsidal line will progrede, while ψ is between $2n\pi + \tau$ and $2(n+1)\pi - \tau$.

To find the angle through which the apsidal line regredes and the period of the regression. Let t', t'' be the values of t, ϖ', ϖ'' the values of ϖ corresponding to the values $2n\pi - \tau$ and $2n\pi + \tau$ of ψ: then

$$(g_1 - g_2) t' + \gamma_1 - \gamma_2 = 2n\pi - \tau,$$
$$(g_1 - g_2) t'' + \gamma_1 - \gamma_2 = 2n\pi + \tau,$$
$$\tan(\varpi' - g_1 t' - \gamma_1) = \frac{M_2 \sin \tau}{M_1 + M_2 \cos \tau},$$
$$\tan(\varpi'' - g_1 t'' - \gamma_1) = \frac{-M_2 \sin \tau}{M_1 + M_2 \cos \tau}.$$

From these equations the values of t', t'', ϖ', ϖ'' may be found, and thus $\varpi' - \varpi''$ the amount of regression will be known. The period of regression

$$= t'' - t' = \frac{2\tau}{g_1 - g_2}.$$

In like manner the amount and period of the progression may be obtained. The latter will be found to be $\dfrac{2(\pi - \tau)}{g_1 - g_2}$.

The period of a complete oscillation will be the sum of the periods of the regression and progression, that is $\dfrac{2\pi}{g_1 - g_2}$, which agrees with the preceding article.

78. The motion of the centre of the instantaneous ellipse in consequence of the secular variations of e and ϖ may be exhibited geometrically as follows.

We have, by Art. 72,

$$e \cos \varpi = M_1 \cos(g_1 t + \gamma_1) + M_2 \cos(g_2 t + \gamma_2),$$
$$e \sin \varpi = M_1 \sin(g_1 t + \gamma_1) + M_2 \sin(g_2 t + \gamma_2).$$

Let a circle be described in the plane of the orbit with its centre S coinciding with that of the sun, and its radius equal to $M_1 a$, where a is the mean distance. Let a point P describe this circle uniformly with a velocity g_1, starting from

O. Again, with centre P and radius equal to $M_2 a$ let another circle be described, and let a point Q describe this circle uniformly with a velocity g_2, starting from C. Let SL be

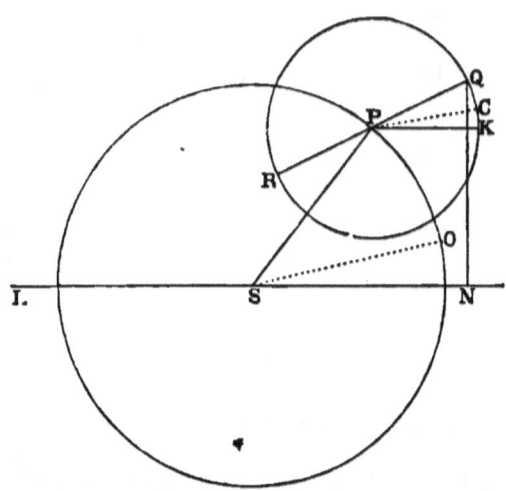

the line from which longitudes are reckoned, and draw PK parallel to it: then if the angle OSN be equal to γ_1, and CPK to γ_2, the angle PSN will be equal to $g_1 t + \gamma_1$, and QPK to $g_2 t + \gamma_2$. Produce QP to meet the circle again in R, and draw QN perpendicular to SN. Then, supposing M_1 and M_2 to be both positive, we have

$$SN = SP \cos PSN + PQ \cos QPK$$
$$= M_1 a \cos (g_1 t + \gamma_1) + M_2 a \cos (g_2 t + \gamma_2)$$
$$= ae \cos \varpi.$$

Similarly, it may be shewn that

$$QN = ae \sin \varpi.$$

Hence, the apse being supposed to move from L in the direction contrary to that of the hands of a watch, Q will be the centre of the instantaneous ellipse.

SECULAR VARIATIONS.

If M_1 be positive and M_2 negative, it may be shewn in like manner that the centre of the ellipse will be R. If M_1, M_2 be both negative, join QS and produce it to Q' so that $SQ' = SQ$: then the centre of the ellipse will be Q'.

A similar construction will of course apply for the motion of the further focus.

79. *To integrate the equations for the inclination and longitude of the node.*

We have (Art. 61) for the planet m

$$\frac{di}{dt} = \frac{m'n a^2 a'}{4\mu} D_1 \tan i'' \sin(\Omega - \Omega'),$$

$$\tan i \frac{d\Omega}{dt} = -\frac{m'n a^2 a'}{4\mu} D_1 \{\tan i - \tan i'' \cos(\Omega - \Omega')\};$$

with similar equations for the planet m'.

To integrate these, assume

$$p = \tan i \sin \Omega, \quad q = \tan i \cos \Omega,$$
$$p' = \tan i'' \sin \Omega', \quad q' = \tan i'' \cos \Omega';$$

therefore $\dfrac{dp}{dt} = \tan i \cos \Omega \dfrac{d\Omega}{dt} + \sin \Omega (1 + \tan^2 i) \dfrac{di}{dt}$.

Substituting the expressions for $\dfrac{d\Omega}{dt}$ and $\dfrac{di}{dt}$, and writing α for $\dfrac{m'n a^2 a'}{4\mu}$, since $\tan^2 i \dfrac{di}{dt}$ being of the third order may be omitted, we have

$$\frac{dp}{dt} = \alpha D_1 (\tan i'' \cos \Omega' - \tan i \cos \Omega)$$
$$= \alpha D_1 (q' - q).$$

Similarly, $\quad \dfrac{dq}{dt} = \alpha D_1 (p - p')$.

Also for the planet m', writing a' for $\frac{mn'a'^2a}{4\mu}$,

$$\frac{dp'}{dt} = a'D_1(q-q'),$$

$$\frac{dq'}{dt} = a'D_1(p'-p).$$

The forms of these equations suggest the following particular integrals:

$$p = N \sin(ht + \delta), \qquad q = N \cos(ht + \delta),$$
$$p' = N'\sin(ht + \delta), \qquad q' = N'\cos(ht + \delta).$$

Substituting these in the differential equations, we obtain from either of the first two,

$$hN = aD_1(N' - N),$$

and from either of the last two,

$$hN' = a'D_1(N - N');$$

eliminating the ratio $N : N'$,

$$(h + aD_1)(h + a'D_1) = aa'D_1^2,$$
or
$$h^2 + (a + a')D_1 h = 0;$$
therefore
$$h = -(a + a')D_1, \text{ or } h = 0.$$

Denote the former by h_1, and let $\delta_1, \delta_2, N_1, N_2, N_1', N_2'$, be the values of δ, N, N' corresponding to $h = h_1$ and $h = 0$. Then $N_2' = N_2$, and the complete solution of the differential equations will be

$$p = N_1 \sin(h_1 t + \delta_1) + N_2 \sin \delta_2,$$
$$q = N_1 \cos(h_1 t + \delta_1) + N_2 \cos \delta_2,$$
$$p' = N_1' \sin(h_1 t + \delta_1) + N_2 \sin \delta_2,$$
$$q' = N_1' \cos(h_1 t + \delta_1) + N_2 \cos \delta_2.$$

Of the constants in these equations, four are arbitrary, and must be determined from the known values of i and Ω at some given epoch.

We have then

$$\tan^2 i = p^2 + q^2 = N_1^2 + N_2^2 + 2N_1N_2 \cos(h_1 t + \delta_1 - \delta_2),$$

$$\tan \Omega = \frac{p}{q} = \frac{N_1 \sin(h_1 t + \delta_1) + N_2 \sin \delta_2}{N_1 \cos(h_1 t + \delta_1) + N_2 \cos \delta_2},$$

with similar equations for i' and Ω'.

Had we considered a system of several planets, we should have obtained a result precisely similar to that of Art. 73.

80. From the form of the expression for $\tan i$, the stability of the inclinations, in the case of two planets, may be inferred. We have

$$\tan^2 i = N_1^2 + N_2^2 + 2N_1N_2 \cos(h_1 t + \delta_1 - \delta_2).$$

Consequently $\tan i$ fluctuates between the limits $N_1 + N_2$ and $N_1 \sim N_2$; and since we know from observation that N_1 and N_2 are very small, it follows that the inclination must always remain very small.

Further, the periods of the changes in the inclinations of the orbits of the two planets are the same, being $\dfrac{2\pi}{\pm h_1}$; and as appears from the equation of Art. 66, the maximum of each inclination will take place at the time of the minimum of the other.

In the case of Jupiter and Saturn, the period is 50673 years; the maximum and minimum inclinations of Jupiter's orbit to the ecliptic are $2°\ 2'\ 30''$ and $1°\ 17'\ 10''$, those of Saturn's orbit $2°\ 32'\ 40''$ and $0°\ 47'$.

81. We now proceed to examine the expression which

has been obtained in Art. 79 for the longitude of the node. We have

$$\tan \Omega = \frac{N_1 \sin(h_1 t + \delta_1) + N_2 \sin \delta_2}{N_1 \cos(h_1 t + \delta_1) + N_2 \cos \delta_2}.$$

The maxima and minima values of Ω, if such exist, will be found by equating $\frac{d\Omega}{dt}$ to zero. Thus

$$\cos(h_1 t + \delta_1 - \delta_2) = -\frac{N_1}{N_2}.$$

If this (disregarding sign) be not greater than unity, the node will oscillate, the period of a complete oscillation being the same as that of the excentricities, viz. $\frac{2\pi}{\pm h}$. But if it be greater than unity, there cannot be any stationary positions, and the node will move continually in one direction.

It may be shewn, as in Art. 76, that the motion of the node will be fastest or slowest whenever the inclination is either a maximum or minimum.

82. *When the line of nodes oscillates, to find the extent and periods of its oscillations.*

It may be shewn as in Art. 77, that if ψ be written for $h_1 t + \delta_1 - \delta_2$, and τ denote the least positive angle whose cosine is $-\frac{N_1}{N_2}$,

$$\tan(\Omega - \delta_2) = \frac{N_1 \sin \psi}{N_2 + N_1 \cos \psi}$$

$$= \frac{-\cos \tau \sin \psi}{1 - \cos \tau \cos \psi},$$

and

$$\tan^2 i \frac{d\Omega}{dt} = h_1 N_1 N_2 (\cos \psi - \cos \tau).$$

Different cases will occur according to the signs of N_1, N_2 and h_1. Suppose N_1, N_2 of like sign, and h_1 negative:

then ψ decreases as t increases, and the line of nodes regredes so long as ψ is between $2n\pi + \tau$ and $2n\pi - \tau$, and progredes so long as ψ is between $2n\pi - \tau$ and $2(n-1)\pi + \tau$.

Let Ω', Ω'' be the values of Ω corresponding to the values $2n\pi + \tau$ and $2n\pi - \tau$ of ψ; then
$$\tan(\Omega' - \delta_2) = -\cot\tau,$$
$$\tan(\Omega'' - \delta_2) = \cot\tau;$$
therefore
$$\Omega' - \delta_2 = m\pi + \tau - \frac{\pi}{2},$$
$$\Omega'' - \delta_2 = m\pi - \left(\tau - \frac{\pi}{2}\right);$$
therefore $\Omega' - \Omega'' = 2\tau - \pi$,

which is the angle through which the line of nodes regredes. Also the period of this regression may be shewn as in Art. 77 to be $\dfrac{2\tau}{-h_1}$. Similarly, the angle through which the line of nodes progredes may be shewn to be $2\tau - \pi$, and the period of the progression $\dfrac{2(\pi - \tau)}{-h_1}$.

The period of a complete oscillation will be the sum of the periods of the regression and progression, that is $\dfrac{2\pi}{-h_1}$, which agrees with the preceding Article.

The remaining cases corresponding to different arrangements of the signs of N_1, N_2 and h_1 may be treated in like manner.

The mean value of Ω is $m\pi + \delta_2$, (this will be found to be the case whatever be the signs of N_1, N_2 and h_1;) and the mean value coincides with the true whenever $\sin\psi = 0$. Since then ψ is the same both for the disturbed and disturbing planet, the nodes of both orbits will arrive simultaneously at their mean positions.

In the case of Jupiter and Saturn N_2 is for each planet numerically less than N_1, so that the node oscillates; the extent of oscillation being 13° 9′ 40″ in Jupiter's orbit, and 31° 56′ 20″ in that of Saturn on either side of their mean position, the ecliptic being taken for the plane of reference, and supposed immoveable.

83. *To shew that the inclination of the orbits of two mutually disturbing planets to each other is approximately constant.*

If γ denote this inclination, we have by Spherical Trigonometry,

$$\cos \gamma = \cos i \cos i'' + \sin i \sin i'' \cos (\Omega - \Omega')$$

$$= \cos i \cos i'' \{1 + \tan i \tan i'' \cos (\Omega - \Omega')\}$$

$$= (1 + \tan^2 i)^{-\frac{1}{2}} (1 + \tan^2 i'')^{-\frac{1}{2}} \{1 + \tan i \tan i'' \cos (\Omega - \Omega')\}$$

$$= 1 - \frac{1}{2} \{\tan^2 i + \tan^2 i'' - 2 \tan i \tan i'' \cos (\Omega - \Omega')\},$$

if we neglect small quantities of orders higher than the second.

Now $\tan^2 i + \tan^2 i'' - 2 \tan i \tan i'' \cos (\Omega - \Omega')$

$$= p^2 + q^2 + p'^2 + q'^2 - 2 (pp' + qq')$$

$$= (p - p')^2 + (q - q')^2$$

$$= (N_1 - N_1')^2;$$

therefore $$1 - \cos \gamma = \frac{1}{2} (N_1 - N_1')^2,$$

or $$\sin \frac{\gamma}{2} = \frac{1}{2} (N_1 - N_1');$$

whence it follows that γ is constant.

84. The equations which give the secular variations

SECULAR VARIATIONS.

of the node and inclination may be interpreted geometrically as follows*.

The equations to be interpreted are

$$p = N_1 \sin(h_1 t + \delta_1) + N_2 \sin \delta_2,$$
$$q = N_1 \cos(h_1 t + \delta_1) + N_2 \cos \delta_2,$$

where
$$p = \tan i \sin \Omega,$$
$$q = \tan i \cos \Omega.$$

Since in the differential equations from which these have been obtained, small quantities of the third order have been neglected, we have to the same order of approximation

$$\left. \begin{array}{l} \sin i \sin \Omega = N_1 \sin(h_1 t + \delta_1) + N_2 \sin \delta_2 \\ \sin i \cos \Omega = N_1 \cos(h_1 t + \delta_1) + N_2 \cos \delta_2 \end{array} \right\} \quad \ldots\ldots (1).$$

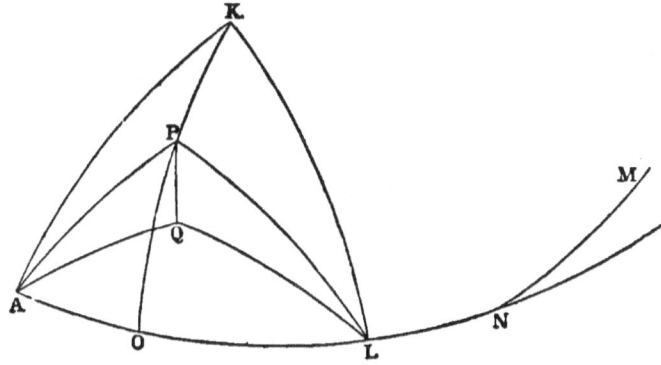

Let a sphere be described with its centre coinciding with that of the sun, and its radius of any magnitude: let the fixed plane of reference cut it in the great circle AL, L being the origin from which longitudes are measured, and LA a

---

* For this elegant geometrical interpretation, the Author is indebted to Mr Freeman, of St John's College. See Appendix, Art. 7.

quadrant. Let another fixed plane inclined at a small angle I to the former cut the sphere in NM, and let $LN = \omega$. Let K, P, Q be the poles of AL, NM, and the plane of the orbit respectively: join $KA, KL, PA, PL, QA, QL, PQ, KP$, and produce the latter to meet AL in O. Then N will be the pole of OK, and ON a quadrant; therefore

$$AO = 90° - OL = LN = \omega.$$

Let $PQ = \rho$, and the angle $QPO = \theta$. Then from the right-angled triangles AOP, LOP, we have

$$\cos AP = \sin I \cos \omega, \qquad \cos LP = \sin I \sin \omega,$$
$$\sin AP \sin APO = \sin \omega, \qquad \sin AP \cos APO = \cos I \cos \omega.$$

Similarly

$$\cos AQ = \sin i \cos \Omega, \qquad \cos LQ = \sin i \sin \Omega.$$

Now $\cos AQ = \cos AP \cos PQ + \sin AP \sin PQ \cos APQ$

$$= \cos AP \cos PQ + \sin AP \sin PQ$$
$$(\cos APO \cos OPQ - \sin APO \sin OPQ),$$

or $\sin i \cos \Omega = \sin I \cos \omega \cos \rho$
$$+ \sin \rho (\cos I \cos \omega \cos \theta - \sin \omega \sin \theta)$$
$$= \sin I \cos \omega \cos \rho + \sin \rho \cos (\theta + \omega),$$

neglecting $\sin^2 \tfrac{1}{2} I \sin \rho.$

Similarly,

$$\sin i \sin \Omega = \sin I \sin \omega \cos \rho + \sin \rho \sin (\theta + \omega).$$

Now these equations will be identical with equations (1), if we suppose

$$\sin \rho = N_1, \qquad \sin I \cos \rho = N_2,$$
$$\omega = \delta_2, \qquad \theta = h_1 t + \delta_1 - \delta_2.$$

SECULAR VARIATIONS.

We have then the following interpretation:—*the normal to the plane of the orbit moves uniformly so as to generate in space a fixed right circular cone.*

85. *To integrate the equation for the longitude of the epoch.*

We have (Art. 61)

$$\frac{d\varepsilon}{dt} = A + A_1 (e^2 - \tan^2 i) + A_2 (e'^2 - \tan^2 i')$$
$$+ A_3 ee' \cos(\varpi - \varpi') + A_4 \tan i \tan i' \cos(\Omega - \Omega').$$

Now from the formulæ of Art. 72, we obtain

$$e^2 = M_1^2 + M_2^2 + 2M_1 M_2 \cos\{(g_1 - g_2) t + \gamma_1 - \gamma_2\},$$
$$e'^2 = M_1'^2 + M_2'^2 + 2M_1' M_2' \cos\{(g_1 - g_2) t + \gamma_1 - \gamma_2\},$$
$$ee' \cos(\varpi - \varpi') = M_1 M_1' + M_2 M_2'$$
$$+ (M_1 M_2' + M_2 M_1') \cos\{(g_1 - g_2) t + \gamma_1 - \gamma_2\}.$$

In like manner, from the formulæ of Art. 79,

$$\tan^2 i = N_1^2 + N_2^2 + 2N_1 N_2 \cos\{h_1 t + \delta_1 - \delta_2\},$$
$$\tan^2 i' = N_1'^2 + N_2'^2 + 2N_1' N_2 \cos\{h_1 t + \delta_1 - \delta_2\},$$
$$\tan i \tan i' \cos(\Omega - \Omega') = N_1 N_1' + N_2^2$$
$$+ N_2 (N_1 + N_1') \cos\{h_1 t + \delta_1 - \delta_2\}.$$

If these values be substituted in the expression for $\frac{d\varepsilon}{dt}$, it takes the form

$$\frac{d\varepsilon}{dt} = Bn + B_1 \cos\{(g_1 - g_2) t + \gamma_1 - \gamma_2\} + B_2 \cos\{h_1 t + \delta_1 - \delta_2\},$$

where Bn, B_1, and B_2 denote certain constants. Integrating, we have

$$\epsilon = \epsilon_0 + Bnt + \frac{B_1}{g_1 - g_2} \sin\{(g_1 - g_2)t + \gamma_1 - \gamma_2\}$$
$$+ \frac{B_2}{h_1} \sin\{h_1 t + \delta_1 - \delta_2\}.$$

We may omit the term Bnt, if we consider it as furnishing a correction on the mean motion n, which thus becomes $(1 + B)n$. With this understanding

$$\epsilon = \epsilon_0 + \frac{B_1}{g_1 - g_2} \sin\{(g_1 - g_2)t + \gamma_1 - \gamma_2\} + \frac{B_2}{h_1} \sin\{h_1 t + \delta_1 - \delta_2\}.$$

If this expression be developed, we may again omit the term involving the first power of t, and consider it as affording a further correction to the mean motion[*]. Thus we shall obtain a series of the form

$$\delta\epsilon = B_3 t^2 + B_4 t^3 + \ldots$$

86. In the Theory of the Planets this inequality is insensible, but in that of the Moon it amounts to upwards of 10 seconds in a century, forming what is termed the *secular acceleration of the Moon's mean motion*. Thus it appears that this inequality does not, as its name would seem to imply, contradict the general theorem of the invariability of the mean motions, since it is due to a variation not of the mean motion, (as we have employed the term in the preceding pages,) but of the epoch. If however, as in the Lunar Theory, the epoch be omitted, any variation in the mean longitude will of necessity be thrown upon the mean motion; only in this case, n will not be given by the equation $n^2 a^3 = \mu$.

87. We have hitherto supposed the planetary motions to be referred to a fixed plane, but have left the particular plane

---

[*] The advantage of thus disposing of these terms arises from the fact that the mean motion as determined by observation is the complete coefficient of t in the expression of the mean longitude.

undetermined. In practice it is usual to take the position of the ecliptic at some given epoch, as for instance the year 1800; but since it is to the true ecliptic that astronomers refer the celestial motions, we will now obtain formulæ for determining relatively to the plane of the earth's orbit, the position of that of any other planet.

Let then m, m' denote the masses of the earth and the planet considered, and suppose the orbits of m and m' but little inclined to each other and to the fixed plane of reference. Let λ, λ' denote the latitudes of points in these orbits corresponding to the same longitude θ_1; then (see fig. to Art. 13)

$$\tan \lambda = \tan i \sin (\theta_1 - \Omega), \quad \tan \lambda' = \tan i' \sin (\theta_1 - \Omega').$$

Now since i and i' are very small, we may replace $\tan \lambda$, $\tan \lambda'$ by λ, λ' respectively: thus

$$\lambda' - \lambda = \tan i' \sin (\theta_1 - \Omega') - \tan i \sin (\theta_1 - \Omega)$$
$$= (\tan i' \cos \Omega' - \tan i \cos \Omega) \sin \theta_1$$
$$\quad - (\tan i' \sin \Omega' - \tan i \sin \Omega) \cos \theta_1,$$

or, with the notation of Art. 79,

$$\lambda' - \lambda = (q' - q) \sin \theta_1 - (p' - p) \cos \theta_1 \ \ldots\ldots\ldots (1).$$

Now let γ denote the inclination, ν the longitude of the node of the orbit of m' relatively to that of m; then approximately

$$\lambda' - \lambda = \tan \gamma \sin (\theta_1 - \nu)$$
$$= \tan \gamma \cos \nu \sin \theta_1 - \tan \gamma \sin \nu \cos \theta_1 \ \ldots\ldots\ldots (2).$$

Hence equating coefficients of $\sin \theta_1$ and $\cos \theta_1$ in equations (1) and (2),

$$\tan \gamma \cos \nu = q' - q, \quad \tan \gamma \sin \nu = p' - p;$$

whence
$$\tan^2 \gamma = (p' - p)^2 + (q' - q)^2,$$

and $$\tan \nu = \frac{p' - q}{q' - q}.$$

These expressions determine the position of the orbit of m' relatively to that of m, when the values of p, p', q, and q' are known. Differentiating them, and neglecting small quantities of orders higher than the second, we obtain

$$\frac{d\gamma}{dt} = \left(\frac{dp'}{dt} - \frac{dp}{dt}\right) \sin \nu + \left(\frac{dq'}{dt} - \frac{dq}{dt}\right) \cos \nu,$$

$$\frac{d\nu}{dt} = \left(\frac{dp'}{dt} - \frac{dp}{dt}\right) \frac{\cos \nu}{\tan \gamma} - \left(\frac{dq'}{dt} - \frac{dq}{dt}\right) \frac{\sin \nu}{\tan \gamma}.$$

If the values of $\frac{dp}{dt}$, $\frac{dq}{dt}$, &c. be substituted, these equations give the variations of γ and ν.

88. For the theory of the invariable plane of the solar system, the reader is referred to Pontécoulant's *Système du Monde*.

CHAPTER VI.

PERIODIC VARIATIONS OF THE ELEMENTS OF THE ORBIT.

89. WE come now to consider the variations produced by the periodical terms of R. These are called *Periodical Variations*, as opposed to the Secular Variations produced by the non-periodic terms. We have seen indeed that the latter are for the most part periodical in form, but in the Planetary Theory, the term Periodical Variations is restricted to those we are about to consider in the present Chapter.

90. We have seen (Art. 46) that the general type of a periodical term is $P \cos \{(pn \pm qn') t + Q\}$, where P is a function of a, e, i; and Q is a function of ϖ, ϵ, Ω. Now such a term will produce a similar term in $\frac{dR}{da}$, $\frac{dR}{de}$, and $\frac{dR}{di}$; but a term of the form $P \sin \{(pn \pm qn') t + Q\}$ in $\frac{dR}{d\varpi}$, $\frac{dR}{d\epsilon}$, and $\frac{dR}{d\Omega}$. If then these be substituted in the equations of Art. 39, they will take the forms

$$\frac{da}{dt} = P_1 \sin \{(pn \pm qn') t + Q\},$$

$$\frac{de}{dt} = P_2 \sin \{(pn \pm qn') t + Q\},$$

$$\frac{d\varpi}{dt} = P_3 \cos\{(pn \pm qn')\,t + Q\},$$

$$\frac{de}{dt} = P_4 \cos\{(pn \pm qn')\,t + Q\},$$

$$\frac{d\Omega}{dt} = P_5 \cos\{(pn \pm qn')\,t + Q\},$$

$$\frac{di}{dt} = P_6 \sin\{(pn \pm qn')\,t + Q\},$$

$$\frac{d^2\zeta}{dt^2} = P_7 \sin\{(pn \pm qn')\,t + Q\},$$

where P_1, P_2, &c. are functions of the elements of the disturbed and disturbing planets, and involve the first power of the disturbing mass.

91. In integrating these equations, we may in general consider the elements which enter in the right-hand members as constant and equal to their values at the epoch from which the time is reckoned*.

Let then a, e, ϖ, &c. denote the values of the elements at epoch, δa, δe, $\delta \varpi$, &c. their periodical variations after an interval t: then integrating the above equations

$$\delta a = -\frac{P_1}{pn \pm qn'} \cos\{(pn \pm qn')\,t + Q\},$$

$$\delta e = -\frac{P_2}{pn \pm qn'} \cos\{(pn \pm qn')\,t + Q\},$$

$$\delta \varpi = \frac{P_3}{pn \pm qn'} \sin\{(pn \pm qn')\,t + Q\},$$

* This is equivalent to neglecting the square of the disturbing force: see Art. 95.

PERIODIC VARIATIONS. 91

$$\delta\epsilon = \frac{P_4}{pn \pm qn'} \sin\{(pn \pm qn')\,t + Q\},$$

$$\delta\Omega = \frac{P_5}{pn \pm qn'} \sin\{(pn \pm qn')\,t + Q\},$$

$$\delta i = -\frac{P_6}{pn \pm qn'} \cos\{(pn \pm qn')\,t + Q\},$$

$$\delta\zeta = -\frac{P_7}{(pn \pm qn')^2} \sin\{(pn \pm qn)\,t + Q\}.$$

Hence it appears that the variations produced by the periodical terms of R are all periodical in form.

92. It will be seen that all the expressions of the last Article involve the divisor $pn \pm qn'$, while $\delta\zeta$ involves the divisor $(pn \pm qn')^2$. If then it should happen that either $pn + qn'$ or $pn \sim qn'$ is very small, a term in R containing $(pn \pm qn')\,t$ in its argument, though of a high order, may have a sensible effect on the elements of the orbit. Now since p and q are either positive integers or zero, $pn + qn'$ cannot be small unless n and n' are small, a case which does not occur with any of the planets: but we have instances in which $pn \sim qn'$ is small*.

Since the period of such inequalities is very great $\Big($being $\frac{2\pi}{pn \sim qn'}\Big)$, they are called *inequalities of long period*, or *long inequalities*.

93. *To select such terms in* R *as will produce the principal inequalities of long period.*

* If in any case the mean motions of two planets were exactly commensurable and in the ratio of p to q, the corresponding term of R, as we have already remarked (Art. 48), would cease to be periodical and would form a part of F, but no instance of this occurs among the planets.

We have seen that the dimension of the principal part of the coefficient of a term containing $(pn \sim qn')\, t$ in its argument is $p \sim q$ (Art. 50); hence if we can find two integers p and q nearly in the ratio of n to n', and having a small difference, the corresponding term of R will produce an important long inequality in the elements of each planet.

In the case of Jupiter and Saturn $n : n' :: 5 : 2$ nearly, and $5 - 2 = 3$; hence there is a long inequality arising from a term in R of the form $P \cos\{(2n - 5n')\, t + Q\}$, the principal part of P being of the third order. This inequality is interesting in an historical point of view, having long baffled the labours of mathematicians and appeared inexplicable on the hypothesis of gravitation. It was at last successfully explained by Laplace.

For the Earth and Venus, $n : n' :: 8 : 13$ nearly, so that there is a long inequality arising from a term in R of the fifth order. The discovery of this inequality is due to the Astronomer Royal.

Finally, in the case of Neptune and Uranus, $n : n' :: 1 : 2$ nearly, hence there is a long inequality arising from a term in R which is of the first order.

94. Between corresponding terms of the long inequalities in the mean motions of two planets, arising from the near commensurability of n and n', there is a simple approximate relation.

Let m, m' be the masses of the two planets, R, R' their disturbing functions: then by Art. 8, considering only the mutual action of m and m', we have

$$R = \frac{m'}{\rho'} - \frac{m'}{r'^3}(xx' + yy' + zz'),$$

PERIODIC VARIATIONS. 93

$$R' = \frac{m}{\rho} - \frac{m}{r^3}(xx' + yy' + zz').$$

We shall distinguish the first and second terms of R and R' as the *symmetrical* and *unsymmetrical* parts respectively, since the co-ordinates of m and m' are involved symmetrically in the former but not in the latter.

Since then the symmetrical parts of R and R' differ only in having m and m' interchanged, if

$$m'M \cos\{(pn - qn')t + Q\}$$

be any term in the symmetrical part of R, that of R' will contain the term

$$mM \cos\{(pn - qn')t + Q\}.$$

Confining our attention to these terms, we have (Art. 39)

$$\frac{d^2\zeta}{dt^2} = -\frac{3n^2a}{\mu}\frac{dR}{d\epsilon} = -\frac{3na}{\mu}\frac{d(R)}{dt}$$

$$= \frac{3n^2ap}{\mu} m'M \sin\{(pn - qn')t + Q\};$$

therefore $\quad \delta\zeta = -\dfrac{3n^2ap}{\mu}\dfrac{m'M}{(pn-qn')^2} \sin\{(pn - qn')t + Q\}.$

Similarly, $\delta\zeta' = \dfrac{3n'^2a'q}{\mu'}\dfrac{mM}{(pn-qn')^2} \sin\{(pn - qn')t + Q\}.$

Hence $\quad \dfrac{\delta\zeta}{\delta\zeta'} = -\dfrac{m'\mu'n^2ap}{m\mu n'^2a'q} = -\dfrac{m'\mu'na}{m\mu n'a'}$

approximately, since qn' is nearly equal to pn; therefore

$$\frac{\delta\zeta}{\delta\zeta'} = -\frac{m'\sqrt{(\mu'a')}}{m\sqrt{(\mu a)}},$$

or since μ' differs from μ by a quantity of the order of the

disturbing force, the square of which we are neglecting, we have
$$\frac{\delta \zeta}{\delta \zeta'} = - \frac{m' \sqrt{a'}}{m \sqrt{a}},$$
the required relation.

The same relation is also approximately true in the case of terms arising from the unsymmetrical parts of R and R'*. For denoting these by R_1 and R_1' respectively, we have

$$R_1 = - m' \left(x \frac{x'}{r'^3} + y \frac{y'}{r'^3} + z \frac{z'}{r'^3} \right),$$

$$R_1' = - m \left(x' \frac{x}{r^3} + y' \frac{y}{r^3} + z' \frac{z}{r^3} \right).$$

Now the equations of motion of the planet m' referred to rectangular axes are

$$\frac{d^2 x'}{dt^2} + \frac{\mu' x'}{r'^3} = \frac{dR'}{dx'}, \text{ \&c.}$$

Hence, the differential coefficients being taken as if the elements were constant†

$$\frac{x'}{r'^3} = - \frac{1}{\mu'} \frac{d^2 x'}{dt^2}, \quad \frac{y'}{r'^3} = - \frac{1}{\mu'} \frac{d^2 y'}{dt^2}, \quad \frac{z'}{r'^3} = - \frac{1}{\mu'} \frac{d^2 z'}{dt^2};$$

therefore
$$R_1 = \frac{m'}{\mu'} \left(x \frac{d^2 x'}{dt^2} + y \frac{d^2 y'}{dt^2} + z \frac{d^2 z'}{dt^2} \right).$$

---

* For the demonstration of this we are mainly indebted to *The Theory of the Long Inequality of Uranus and Neptune:* an essay which obtained the Adams Prize for the year 1850. By R. Pierson, M.A.

† This is simply an analytical artifice: we merely assert that *if* the differential coefficients be so taken, *then* $\frac{x'}{r'^3}$, &c., and therefore R_1 and R_1' will take the above forms.

Similarly, $R_1' = \dfrac{m}{\mu}\left(x'\dfrac{d^2x}{dt^2} + y'\dfrac{d^2y}{dt^2} + z'\dfrac{d^2z}{dt^2}\right).$

Now any term in R_1 containing $(pn - qn')t$ in its argument can arise only from the combination of terms in x, y, and z, containing pnt in their arguments with terms in $\dfrac{d^2x'}{dt^2}$, $\dfrac{d^2y'}{dt^2}$, and $\dfrac{d^2z'}{dt^2}$ containing $qn't$. Suppose then x and x' when developed in terms of t and the elements to contain respectively the terms

$$L\cos(pnt + l),\quad L'\cos(qn't + l').$$

Hence the product $x\dfrac{d^2x'}{dt^2}$ will contain the term

$$-\tfrac{1}{2}LL'q^2n'^2\cos\{(pn - qn')t + l - l'\},$$

and the product $x'\dfrac{d^2x}{dt^2}$ the term

$$-\tfrac{1}{2}LL'p^2n^2\cos\{(pn - qn')t + l - l'\}:$$

the coefficients are in the ratio $q^2n'^2$ to p^2n^2. Similarly, the coefficients of the same cosine in $y\dfrac{d^2y'}{dt^2}$ and $z\dfrac{d^2z'}{dt^2}$ are to those in $y'\dfrac{d^2y}{dt^2}$ and $z'\dfrac{d^2z}{dt^2}$ in the same ratio.

Hence if $\quad\dfrac{m'}{\mu}Mq^2n'^2\cos\{(pn - qn')t + Q\}$

be any term in R_1, then R_1' will contain the term

$$\dfrac{m}{\mu}Mp^2n^2\cos\{(pn - qn')t + Q\}.$$

Confining our attention to these terms, we have

$$\delta\zeta = -\frac{3n^2 ap}{\mu\mu'} \frac{m'Mq^2 n'^2}{(pn-qn')^2} \sin\{(pn-qn')t + Q\},$$

$$\delta\zeta' = \frac{3n'^2 a'q}{\mu\mu'} \frac{mMp^2 n^2}{(pn-qn')^2} \sin\{(pn-qn')t + Q\}.$$

Hence
$$\frac{\delta\zeta}{\delta\zeta'} = -\frac{m'aq}{ma'p} = -\frac{m'na}{mn'a'}, \text{ nearly},$$

$$= -\frac{m'\sqrt{a'}}{m\sqrt{a}},$$

the square of the disturbing force being neglected.

By means of this relation, when one of the long inequalities is known, the other may be calculated: it may be used as a formula of verification.

95. We have remarked that in integrating the equations of Art. 90, we may in general consider the elements which enter in the right-hand members as constant and equal to their values at the epoch from which the time is measured. In the case, however, of inequalities whose periods are very long, the secular variations of the elements in the interval produce a sensible effect. In order to take account of these, we may integrate our equations by parts, considering the elements variable; and then substitute their values as calculated by the method of Art. 62. For example, consider the equation

$$\frac{d^2\zeta}{dt^2} = P\sin\{(pn-qn')t + Q\}$$

$$= P\sin\lambda, \text{ suppose.}$$

Integrating by parts, and remembering that n is constant with regard to secular variations, we have

PERIODIC VARIATIONS.

$$\frac{d\zeta}{dt} = -\frac{P}{pn-qn'}\cos\lambda + \frac{1}{(pn-qn')^2}\frac{dP}{dt}\sin\lambda$$
$$+\frac{1}{(pn-qn')^3}\frac{d^2P}{dt^2}\cos\lambda - \ldots;$$

therefore $\delta\zeta = -\dfrac{P}{(pn-qn')^2}\sin\lambda - \dfrac{1}{(pn-qn')^3}\dfrac{dP}{dt}\cos\lambda$
$$+\frac{1}{(pn-qn')^4}\frac{d^2P}{dt^2}\sin\lambda + \ldots$$
$$-\frac{1}{(pn-qn')^3}\frac{dP}{dt}\cos\lambda + \frac{1}{(pn-qn')^4}\frac{d^2P}{dt^2}\sin\lambda + \ldots$$
$$+\frac{1}{(pn-qn')^4}\frac{d^2P}{dt^2}\sin\lambda + \ldots$$
$$=\left\{-\frac{P}{(pn-qn')^2} + \frac{3}{(pn-qn')^4}\frac{d^2P}{dt^2} - \ldots\right\}\sin\lambda$$
$$+\left\{-\frac{2}{(pn-qn')^3}\frac{dP}{dt} - \ldots\right\}\cos\lambda.$$

In this equation P, $\dfrac{dP}{dt}$, $\dfrac{d^2P}{dt^2}$, &c. are functions of the elements; their values may be calculated by the formulæ of Art. 62. It may be noticed that P is of the first order, $\dfrac{dP}{dt}$ of the second, and $\dfrac{d^2P}{dt^2}$ of the third of the disturbing force: for $\dfrac{dP}{dt}$, being found from P by differentiation, will involve the differential coefficients of the elements, which are themselves of the first order; and similarly for $\dfrac{d^2P}{dt^2}$.

96. Having now completed our account of the methods

C. P. T.

of treating the secular and periodic variations of the elements of the orbit, we will say a few words on the distinction between them. In the first place we may observe that the periodic variations involve the mean longitude of the disturbed and disturbing planets, and therefore depend chiefly upon the configuration of the planetary system. On the contrary the secular variations depend solely upon the values of the elements. The latter class of variations take place with extreme slowness, so that if these only existed, a considerable time must elapse before the deviation of the planet from elliptic motion became appreciable. On the other hand, the periodic variations (such at least as are rapidly periodic) "are in their nature transient and temporary: they disappear in short periods, and leave no trace. The planet is temporarily drawn from its orbit (its slowly varying orbit), but forthwith returns to it, to deviate presently as much the other way, while the varied orbit accommodates and adjusts itself to the average of these excursions on either side of it; and thus continues to present, for a succession of indefinite ages, a kind of medium picture of all that the planet has been doing in their lapse, in which the expression and character is preserved; but the individual features are merged and lost*." On this account it is convenient to suppose the planet to move in an ellipse, the elements of which are corrected for secular variations only, and to take account of the periodic variations by applying small corrections to the radius vector and longitude as calculated from the elliptic formulæ.

97. We will accordingly shew how by means of the periodic variations of the elements, the corresponding inequalities in the radius vector and longitude may be calculated. If we take for our plane of reference the position of

* Herschel's *Outlines of Astronomy*, 5th edit. Art. 656.

the plane of the orbit of the disturbed planet at the epoch from which the time is reckoned, the inclination will be of the order of the disturbing force, and therefore if we neglect the square of the latter, we may also neglect the square of the former.

98. *To calculate the periodic variations in radius vector.*

Let δa, δe, $\delta \varpi$, &c. denote the periodic variations in a, e, ϖ, &c., and let δr be the corresponding variation in r; then

$$\delta r = \frac{dr}{da}\delta a + \frac{dr}{de}\delta e + \frac{dr}{d\varpi}\delta\varpi + \frac{dr}{d\zeta}\delta\zeta + \frac{dr}{d\epsilon}\delta\epsilon,$$

in which the square of the disturbing force is neglected, since this would be introduced by the squares and products of δa, δe, &c. The values of δa, δe, &c. have been found in Art. 91, those of $\frac{dr}{da}$, $\frac{dr}{de}$, &c. may be obtained from the equation (Art. 40)

$$r = a\left\{1 + \frac{1}{2}e^2 - e\cos(\zeta + \epsilon - \varpi) - \frac{1}{2}e^2\cos 2(\zeta + \epsilon - \varpi) - \ldots\right\}.$$

99. *To calculate the periodic variations in longitude.*

These might be found in the same manner as the variations in radius vector, but they may also be deduced from them: we proceed to obtain a formula for this purpose.

We have $\quad\dfrac{d\theta_0}{dt} = \dfrac{h}{r^2}$, (Art. 22),

and $\quad\theta - \theta_0 = \Omega - \Omega_0$;

therefore $\quad\dfrac{d\theta}{dt} = \dfrac{h}{r^2} + (1 - \cos i)\dfrac{d\Omega}{dt}\quad$ (see Art. 29)

$$= \frac{h}{r^2},$$

since $(1-\cos i)\dfrac{d\Omega}{dt}$ being of the order of the square of the disturbing force may be neglected.

Let δr, $\delta\theta$, and δh be corresponding variations in r, θ and h; then

$$\frac{d(\theta+\delta\theta)}{dt}=\frac{h+\delta h}{(r+\delta r)^2},$$

or $\quad\dfrac{d\theta}{dt}+\dfrac{d\delta\theta}{dt}=\dfrac{h}{r^2}\left(1+\dfrac{\delta h}{h}\right)\left(1+\dfrac{\delta r}{r}\right)^{-2}$

$$=\frac{h}{r^2}+\frac{\delta h}{r^2}-\frac{2h\,\delta r}{r^3},$$

neglecting the square of the disturbing force; therefore

$$\frac{d\delta\theta}{dt}=\frac{\delta h}{r^2}-\frac{2h\,\delta r}{r^3},$$

which gives the variations in longitude. The value of δh may be found from the formula

$$\frac{dh}{dt}=\frac{dR}{d\epsilon}+\frac{dR}{d\varpi}.$$

For the periodic variations in latitude, we refer to Pontécoulant's *Système du Monde*, Tome I. p. 492.

100. As an example of the processes of this chapter, we will calculate the variations in radius vector and longitude due to the term $m'Me\cos\{(n-2n')t+\epsilon-2\epsilon'+\varpi\}$ in R.

Considering this term only, we have

$$R = m'Me\cos\{(n-2n')t+\epsilon-2\epsilon'+\varpi\}$$
$$= m'Me\cos\lambda,\text{ suppose.}$$

Hence $\quad\dfrac{dR}{da}=m'\dfrac{dM}{da}e\cos\lambda,\qquad\dfrac{dR}{de}=m'M\cos\lambda,$

$\qquad\dfrac{dR}{d\epsilon}=-m'Me\sin\lambda,\qquad\dfrac{dR}{d\varpi}=-m'Me\sin\lambda,$

$\qquad\dfrac{dR}{di}=0,\qquad\qquad\qquad\dfrac{dR}{d\Omega}=0.$

PERIODIC VARIATIONS. 101

Substituting these in the formulæ of Art. 39, and neglecting small quantities of orders higher than the first, we have

$$\frac{da}{dt} = -\frac{2na^2}{\mu} m'Me \sin \lambda,$$

$$\frac{de}{dt} = \frac{na}{\mu} m'M \sin \lambda,$$

$$e\frac{d\varpi}{dt} = \frac{na}{\mu} m'M \cos \lambda,$$

$$\frac{d\epsilon}{dt} = -\frac{2na^2}{\mu} m'\frac{dM}{da} e \cos \lambda + \frac{1}{2}\frac{na}{\mu} em'M \cos \lambda,$$

$$\frac{d^2\zeta}{dt^2} = \frac{3n^2a}{\mu} m'Me \sin \lambda.$$

By integration we have

$$\delta a = \frac{2m'M}{\mu} \frac{na^2 e}{n - 2n'} \cos \lambda,$$

$$\delta e = -\frac{m'M}{\mu} \frac{na}{n - 2n'} \cos \lambda,$$

$$e\delta\varpi = \frac{m'M}{\mu} \frac{na}{n - 2n'} \sin \lambda,$$

$$\delta\epsilon = \left(\frac{1}{2}\frac{m'M}{\mu} - \frac{2m'a}{\mu}\frac{dM}{da}\right) \frac{nae}{n - 2n'} \sin \lambda,$$

$$\delta\zeta = -\frac{3m'M}{\mu} \frac{n^2ae}{(n - 2n')^2} \sin \lambda.$$

Also

$$r = a\left\{1 + \frac{1}{2}e^2 - e\cos(\zeta + \epsilon - \varpi) - \frac{1}{2}e^2\cos 2(\zeta + \epsilon - \varpi) - \ldots\right\};$$

therefore, small quantities of orders higher than the first being neglected,

$$\frac{dr}{da} = 1 - e \cos(\zeta + \epsilon - \varpi),$$

$$\frac{dr}{de} = a\{e - \cos(\zeta + \epsilon - \varpi) - e \cos 2(\zeta + \epsilon - \varpi)\},$$

$$\frac{dr}{d\varpi} = -a\{e \sin(\zeta + \epsilon - \varpi) + e^2 \sin 2(\zeta + \epsilon - \varpi)\},$$

$$\frac{dr}{d\zeta} = ae \sin(\zeta + \epsilon - \varpi),$$

$$\frac{dr}{d\epsilon} = ae \sin(\zeta + \epsilon - \varpi).$$

Now $\delta r = \dfrac{dr}{da} \delta a + \dfrac{dr}{de} \delta e + \dfrac{dr}{d\varpi} \delta \varpi + \dfrac{dr}{d\zeta} \delta \zeta + \dfrac{dr}{d\epsilon} \delta \epsilon$

$$= \frac{2m'M}{\mu} \frac{na^2 e}{n - 2n'} \cos \lambda$$

$$- \frac{m'M}{\mu} \frac{na^2}{n - 2n'} \cos \lambda \{e - \cos(\zeta + \epsilon - \varpi) - e \cos 2(\zeta + \epsilon - \varpi)\}$$

$$- \frac{m'M}{\mu} \frac{na^2}{n - 2n'} \sin \lambda \{\sin(\zeta + \epsilon - \varpi) + e \sin 2(\zeta + \epsilon - \varpi)\}.$$

Since we are neglecting the square of the disturbing force, the elements in this equation may be considered as constant, and therefore nt written for ζ: we have then, restoring to λ its value

$$\delta r = \frac{m'M}{\mu} \frac{na^2}{n - 2n'} \cos 2\{(n - n')t + \epsilon - \epsilon'\}$$

$$+ \frac{m'M}{\mu} \frac{na^2 e}{n - 2n'} \cos \{(n - 2n')t + \epsilon - 2\epsilon' + \varpi\}$$

PERIODIC VARIATIONS. 103

$$+ \frac{m'M}{\mu} \frac{na^2e}{n-2n'} \cos\{(3n-2n')t + 3\epsilon - 2\epsilon' - \varpi\},$$

which is the variation in radius vector.

101. To calculate the variation in longitude, we shall employ the equation

$$\frac{d\,\delta\theta}{dt} = \frac{\delta h}{r^2} - \frac{2h\delta r}{r^3}.$$

Now $\quad \dfrac{dh}{dt} = \dfrac{dR}{d\epsilon} + \dfrac{dR}{d\varpi} = -2m'Me\sin\lambda;$

therefore $\quad \delta h = \dfrac{2m'Me}{n-2n'} \cos\lambda;$

therefore $\quad \dfrac{\delta h}{r^2} = \dfrac{2m'Me}{a^2(n-2n')} \cos\lambda$

$$= \frac{2m'M}{\mu} \frac{n^2 ae}{n-2n'} \cos\{(n-2n')t + \epsilon - 2\epsilon' + \varpi\}.$$

Also $\quad \dfrac{h\delta r}{r^3} = \dfrac{h\delta r}{a^3} \{1 + 3e\cos(nt + \epsilon - \varpi) + \ldots\}$

$$= \frac{mM'}{\mu} \frac{n^2 a}{n-2n'} \cos 2\{(n-n')t + \epsilon - \epsilon'\}$$

$$+ \frac{5}{2} \frac{mM}{\mu} \frac{n^2 ae}{n-2n'} \cos\{(n-2n')t + \epsilon - 2\epsilon' + \varpi\}$$

$$+ \frac{5}{2} \frac{m'M}{\mu} \frac{n^2 ae}{n-2n'} \cos\{(3n-2n')t + 3\epsilon - 2\epsilon' - \varpi\}.$$

Hence by substitution

$$\frac{d\,\delta\theta}{dt} = -\frac{2m'M}{\mu} \frac{n^2 a}{n-2n'} \cos 2\{(n-n')t + \epsilon - \epsilon'\}$$

$$-\frac{3m'M}{\mu}\frac{n^2ae}{n-2n'}\cos\{(n-2n')t+\epsilon-2\epsilon'+\varpi\}$$

$$-\frac{5m'M}{\mu}\frac{n^2ae}{n-2n'}\cos\{(3n-2n')t+3\epsilon-2\epsilon'-\varpi\}.$$

By integration

$$\delta\theta = -\frac{m'M}{\mu}\frac{n^2a}{(n-2n')(n-n')}\sin 2\{(n-n')t+\epsilon-\epsilon'\}$$

$$-\frac{3m'M}{\mu}\frac{n^2ae}{(n-2n')^2}\sin\{(n-2n')t+\epsilon-2\epsilon'+\varpi\}$$

$$-\frac{5m'M}{\mu}\frac{n^2ae}{(n-2n')(3n-2n')}\sin\{(3n-2n')t+3\epsilon-2\epsilon'-\varpi\},$$

which is the variation in longitude.

In the case of Uranus and Neptune, since $n : n'$ nearly as $2 : 1$, the term we have been considering is important in the theory of the long inequality.

CHAPTER VII.

DIRECT METHOD OF CALCULATING THE INEQUALITIES IN RADIUS VECTOR, LONGITUDE, AND LATITUDE.

102. In the calculation of the planetary inequalities, we have hitherto employed exclusively the method of the Variation of Elements, but there is another method of solving the problem, which demands our attention. It consists in obtaining equations for calculating the inequalities in radius vector, longitude, and latitude directly from the equations of motion. The two methods are sometimes distinguished, the former as that of *Lagrange*, the latter as that of *Laplace*. In practice both are employed, that of Lagrange chiefly for secular, that of Laplace for periodic variations: but the method of Lagrange may also be advantageously employed in the calculation of long inequalities. We proceed then to explain the method of Laplace.

103. If r_1, θ_1, and z denote the projected radius vector, longitude, and distance from the plane of reference, of the planet, we have (see Art. 9) the equations of motion

$$\frac{d^2 r_1}{dt^2} - r_1 \left(\frac{d\theta_1}{dt}\right)^2 = -\frac{\mu r_1}{r^3} + \frac{dR}{dr_1},$$

$$\frac{d}{dt}\left(r_1^2 \frac{d\theta_1}{dt}\right) = \frac{dR}{d\theta_1},$$

$$\frac{d^2 z}{dt^2} = -\frac{\mu z}{r^3} + \frac{dR}{dz}.$$

If we take for the fixed plane of reference the position of the plane of the orbit at the epoch from which the time is measured, the inclination (as we have remarked in Art. 97) will be the order of the disturbing force, the square of which will be neglected. Now it will be seen on referring to Art. 42, that r_1 and θ_1 differ from r and θ by quantities depending upon the square of the inclination: hence in the above equations, we may replace r_1 and θ_1 by r and θ respectively. Also if λ denote the latitude of the planet, we have

$$z = r \sin \lambda.$$

Hence our equations of motion become

$$\frac{d^2 r}{dt^2} - r\left(\frac{d\theta}{dt}\right)^2 = -\frac{\mu}{r^2} + \frac{dR}{dr} \quad \ldots\ldots\ldots\ldots\ldots (1),$$

$$\frac{d}{dt}\left(r^2 \frac{d\theta}{dt}\right) = \frac{dR}{d\theta} \quad \ldots\ldots\ldots\ldots\ldots\ldots\ldots (2),$$

$$\frac{d^2 (r \sin \lambda)}{dt^2} = -\frac{\mu}{r^2} \sin \lambda + \frac{dR}{dz} \quad \ldots\ldots\ldots (3).$$

104. As a first approximation, let values of r, θ and λ be obtained from these equations by neglecting the disturbing force, and let $r + \delta r$, $\theta + \delta \theta$, $\lambda + \delta \lambda$ denote the true values of these co-ordinates; then δr, $\delta \theta$ and $\delta \lambda$ will be very small quantities, of the order of the disturbing force: they are termed the *perturbations* in radius vector, longitude, and latitude. We proceed to investigate equations by means of which these quantities may be determined.

105. *To obtain the equation for the perturbation in radius vector.*

From equations (1) and (2) of Art. 103, we obtain

DIRECT METHOD OF CALCULATION.

$$\left(\frac{dr}{dt}\right)^2 + r^2\left(\frac{d\theta}{dt}\right)^2 = \frac{2\mu}{r} + 2\int\left(\frac{dR}{dr}\frac{dr}{dt} + \frac{dR}{d\theta}\frac{d\theta}{dt}\right) + C$$

$$= \frac{2\mu}{r} + 2\int\frac{d(R)}{dt}dt + C \quad \text{............(4)}.$$

Multiply (1) by r and add it to (4): thus

$$r\frac{d^2r}{dt^2} + \left(\frac{dr}{dt}\right)^2 = \frac{\mu}{r} + r\frac{dR}{dr} + 2\int\frac{d(R)}{dt}dt + C,$$

i.e. $$\frac{d^2(r^2)}{dt^2} = \frac{2\mu}{r} + 2r\frac{dR}{dr} + 4\int\frac{d(R)}{dt}dt + 2C \quad \text{........(5)}.$$

If the disturbing force be neglected, this equation becomes

$$\frac{d^2(r^2)}{dt^2} = \frac{2\mu}{r} + 2C \quad \text{....................(6)}.$$

Let a value of r be obtained from this equation, and let $r + \delta r$ denote the true radius vector: then if we agree to neglect the square of the disturbing force in our next approximation, it will be sufficient to write $r + \delta r$ for r in those terms of (5) which do not involve the disturbing force: also since δr is itself of the order of the disturbing force, its square may be neglected. Hence from (5)

$$\frac{d^2(r+\delta r)^2}{dt^2} = \frac{2\mu}{r+\delta r} + 2r\frac{dR}{dr} + 4\int\frac{d(R)}{dt}dt + 2C;$$

therefore $$\frac{d^2(r^2)}{dt^2} + 2\frac{d^2(r\delta r)}{dt^2} = \frac{2\mu}{r} - \frac{2\mu}{r^2}\delta r + 2r\frac{dR}{dr}$$

$$+ 4\int\frac{d(R)}{dt}dt + 2C;$$

hence by (6),

$$\frac{d^2(r\delta r)}{dt^2} + \frac{\mu}{r^3}(r\delta r) = r\frac{dR}{dr} + 2\int\frac{d(R)}{dt}dt,$$

which is the equation for the perturbation in radius vector. We may express the right-hand member in a more convenient form, for since

$$r = a(1 + u),$$

$$\frac{dR}{da} = \frac{dR}{dr}\frac{dr}{da} = (1+u)\frac{dR}{dr};$$

therefore
$$r\frac{dR}{dr} = a\frac{dR}{da}:$$

also
$$\frac{d(R)}{dt} = n\frac{dR}{d\epsilon}.$$

Hence our equation becomes

$$\frac{d^2(r\delta r)}{dt^2} + \frac{\mu}{r^3}(r\delta r) = a\frac{dR}{da} + 2n\int\frac{dR}{d\epsilon}dt.$$

106. *To obtain the equation for the perturbation in longitude.*

We have from equation (1) of Art. 103,

$$\left(\frac{d\theta}{dt}\right)^2 = \frac{1}{r}\frac{d^2r}{dt^2} + \frac{\mu}{r^3} - \frac{1}{r}\frac{dR}{dr}.$$

As before, let a value of θ be obtained from this equation, the disturbing force being neglected, and let $\theta + \delta\theta$ denote the true value of the longitude: then writing $r + \delta r$ for r and $\theta + \delta\theta$ for θ, and neglecting the square of the disturbing force, we have

$$2\frac{d\theta}{dt}\frac{d\delta\theta}{dt} = \frac{1}{r}\frac{d^2\delta r}{dt^2} - \frac{\delta r}{r^2}\frac{d^2r}{dt^2} - \frac{3\mu}{r^4}\delta r - \frac{1}{r}\frac{dR}{dr}:$$

but
$$r^2\frac{d\theta}{dt} = h, \quad r\frac{dR}{dr} = a\frac{dR}{da};$$

therefore $2h\dfrac{d\delta\theta}{dt} = r\dfrac{d^2\delta r}{dt^2} - \delta r\dfrac{d^2r}{dt^2} - \dfrac{3\mu}{r^3}\delta r - a\dfrac{dR}{da}$

$= \dfrac{d}{dt}\left(r\dfrac{d\delta r}{dt} - \delta r\dfrac{dr}{dt}\right) - \dfrac{3\mu}{r^3}.r\delta r - a\dfrac{dR}{da}.$

This equation will become integrable if we eliminate the term $-\dfrac{3\mu}{r^3}.r\delta r$ by means of the equation for the perturbation in radius vector. We have from that equation

$$0 = 3\dfrac{d^2(r\delta r)}{dt^2} + \dfrac{3\mu}{r^3}.r\delta r - 3a\dfrac{dR}{da} - 6n\int\dfrac{dR}{d\epsilon}dt;$$

therefore by addition,

$$2h\dfrac{d\delta\theta}{dt} = 3\dfrac{d^2(r\delta r)}{dt^2} + \dfrac{d}{dt}\left(r\dfrac{d\delta r}{dt} - \delta r\dfrac{dr}{dt}\right) - 4a\dfrac{dR}{da}$$
$$- 6n\int\dfrac{dR}{d\epsilon}dt;$$

therefore $2h\delta\theta = 3\dfrac{d(r\delta r)}{dt} + r\dfrac{d\delta r}{dt} - \delta r\dfrac{dr}{dt} - 4a\int\dfrac{dR}{da}dt$
$$- 6n\iint\dfrac{dR}{d\epsilon}dt^2,$$

$= 4\dfrac{d(r\delta r)}{dt} - 2\delta r\dfrac{dr}{dt} - 4a\int\dfrac{dR}{da}dt - 6n\iint\dfrac{dR}{d\epsilon}dt^2,$

or $h\delta\theta = 2\dfrac{d(r\delta r)}{dt} - \delta r\dfrac{dr}{dt} - 2a\int\dfrac{dR}{da}dt - 3n\iint\dfrac{dR}{d\epsilon}dt^2,$

which determines the perturbation in longitude, when that in radius vector is known.

107. *To obtain the equation for the perturbation in latitude.*

From equation (3) of Art. 103, we have

$$\dfrac{d^2(r\sin\lambda)}{dt^2} + \dfrac{\mu(r\sin\lambda)}{r^3} = \dfrac{dR}{dz}.$$

Since the plane of reference is supposed to coincide with the position of the plane of the orbit at the epoch from which the time is reckoned, we have at the epoch $\lambda = 0$: hence, denoting by $\delta\lambda$ the latitude at time t, our equation becomes

$$\frac{d^2(r\delta\lambda)}{dt^2} + \frac{\mu}{r^3}(r\delta\lambda) = \frac{dR}{dz},$$

which is similar in form to the equation for the perturbation in radius vector.

108. *To integrate the equation for the perturbation in radius vector.*

The equation is (Art. 105),

$$\frac{d^2(r\delta r)}{dt^2} + \frac{\mu}{r^3}(r\delta r) = a\frac{dR}{da} + 2n\int \frac{dR}{d\epsilon} dt.$$

Let us consider a term in R of the form

$$P \cos\{(pn - qn')t + Q\},$$

where P is a function of the mean distances, excentricities, and inclinations, and Q of the longitudes of the perihelia, nodes, and epochs: then uniting this term with the non-periodic part of R, which we have denoted by F, we have

$$R = F + P\cos\{(pn - qn')t + Q\};$$

therefore $\quad \dfrac{dR}{da} = \dfrac{dF}{da} + \dfrac{dP}{da}\cos\{(pn - qn')t + Q\},$

$$\frac{dR}{d\epsilon} = -P\frac{dQ}{d\epsilon}\sin\{(pn - qn')t + Q\},$$

since F does not contain ϵ; therefore

$$n\int \frac{dR}{d\epsilon} dt = \frac{nP\dfrac{dQ}{d\epsilon}}{pn - qn'}\cos\{(pn - qn')t + Q\} + m'g,$$

where g is an arbitrary constant, the value of which may be any whatever. We might omit it, and consider it as included in the constant C of Art. 105, but the advantage of retaining it will be seen hereafter.

Hence
$$a\frac{dR}{da} + 2n\int \frac{dR}{d\epsilon}\, dt = 2m'g + a\frac{dF}{da}$$
$$+ \left\{a\frac{dP}{da} + \frac{2nP\frac{dQ}{d\epsilon}}{pn - qn'}\right\}\cos\{(pn - qn')t + Q\}$$
$$= 2m'g + a\frac{dF}{da} + P_1 \cos\{(pn - qn')t + Q\},$$

suppose, where
$$P_1 = a\frac{dP}{da} + \frac{2nP\frac{dQ}{d\epsilon}}{pn - qn'}.$$

Again (see Art. 13),
$$r = a\left\{1 + \frac{1}{2}e^2 - e\cos(nt + \epsilon - \varpi)\right.$$
$$\left. - \frac{1}{2}e^2 \cos 2(nt + \epsilon - \varpi) - \ldots\right\},$$

therefore
$$\frac{\mu}{r^3} = \frac{\mu}{a^3}\{1 + 3e\cos(nt + \epsilon - \varpi) + \ldots\},$$
$$= n^2\{1 + 3e\cos(nt + \epsilon - \varpi) + \ldots\},$$

since $n^2 a^3 = \mu.$

Hence, by substitution, the equation for the perturbation in radius vector becomes
$$\frac{d^2(r\delta r)}{dt^2} + n^2 \cdot r\delta r = 2m'g + a\frac{dF}{da} + P_1\cos\{(pn - qn')t + Q\}$$
$$- n^2\, r\delta r\{3e\cos(nt + \epsilon - \varpi) + \ldots\}.$$

109. This equation must be solved by successive approximation, as in the Lunar Theory. By omitting all

small quantities, we obtain a first approximation to the value of $r\delta r$; this being substituted in the second member, and small quantities of orders higher than the first neglected, we obtain a second approximation, which will be correct to the first order. In like manner, a third, and higher approximations may be obtained.

On referring to Art. 59, it will be seen that small quantities of the second order being neglected,

$$F = \frac{1}{2} m' C_0;$$

hence, neglecting all small quantities, the equation of the preceding Article becomes

$$\frac{d^2(r\delta r)}{dt^2} + n^2 \cdot r\delta r = 2m'g + \frac{m'}{2} a \frac{dC_0}{da} + P_1 \cos\{(pn - qn')t + Q\}.$$

The integral of this equation is

$$r\delta r = \frac{1}{n^2}\left(2m'g + \frac{m'}{2} a \frac{dC_0}{da}\right)$$
$$+ \frac{P_1}{n^2 - (pn - qn')^2} \cos\{(pn - qn')t + Q\}$$
$$+ A \cos(nt - B),$$

where A and B are arbitrary constants. Since, if all small quantities be neglected, $r = a$, we have as a first approximation,

$$\delta r = \frac{1}{n^2 a}\left(2m'g + \frac{m'}{2} a \frac{dC_0}{da}\right)$$
$$+ \frac{P_1}{a\{n^2 - (pn - qn')^2\}} \cos\{(pn - qn')t + Q\}$$
$$+ \frac{A}{a} \cos(nt - B).$$

110. We may, however, omit the last term: for, considering this only, the radius vector of the planet becomes

$$a\left\{1 - e\cos(nt + \epsilon - \varpi) + \frac{A}{a^2}\cos(nt - B) + \ldots\right\}$$

$$= a\left[1 - \{e\cos(\epsilon - \varpi) - \frac{A}{a^2}\cos B\}\cos nt\right.$$

$$\left.+ \{e\sin(\epsilon - \varpi) + \frac{A}{a^2}\sin B\}\sin nt + \ldots\right]$$

$$= a\{1 - e_1\cos(nt + \epsilon - \varpi_1) + \ldots\},$$

if
$$e_1\cos(\epsilon - \varpi_1) = e\cos(\epsilon - \varpi) - \frac{A}{a^2}\cos B,$$

$$e_1\sin(\epsilon - \varpi_1) = e\sin(\epsilon - \varpi) + \frac{A}{a^2}\sin B,$$

from which e_1 and ϖ_1 may be determined.

Now since the ellipse upon which our approximations are based, has been obtained by neglecting the disturbing force, we may in the elliptic formulæ replace e and ϖ by e_1 and ϖ_1 respectively, since they differ by quantities of the order of the disturbing force. If this be done, our first approximation becomes

$$\delta r = \frac{1}{n^2 a}\left(2m'g + \frac{m'}{2}a\frac{dC_0}{da}\right)$$

$$+ \frac{P_1}{a\{n^2 - (pn - qn')^2\}}\cos\{(pn - qn')t + Q\}.$$

111. In order to obtain a second approximation, this value must be substituted for δr in the right-hand member of the equation of Art. 108. Also since the square of the disturbing force is neglected, we may write e_1 and ϖ_1 for e and ϖ in this equation. We will write for brevity

$$\delta r = L + P_2\cos\{(pn - qn')t + Q\}.$$

Substituting this in the equation of Art. 108, and omitting those terms which have produced the first approximation*, we have

$$\frac{d^2 . r\delta r}{dt^2} + n^2 . r\delta r$$

$$= - 3n^2 e_1 \cos(nt + \epsilon - \varpi_1) [L + P_2 \cos\{(pn - qn')t + Q\}]$$

$$= - 3n^2 e_1 L \cos(nt + \epsilon - \varpi_1)$$

$$- \frac{3}{2} n^2 e_1 P_2 \cos[\{(p+1)n - qn'\}t + Q + \epsilon - \varpi_1]$$

$$- \frac{3}{2} n^2 e_1 P_2 \cos[\{(p-1)n - qn'\}t + Q - \epsilon + \varpi_1].$$

112. On the form of this equation, we have an important remark to make. In consequence of the term

$$- 3n^2 e_1 L \cos(nt + \epsilon - \varpi_1),$$

its integral will contain the term

$$\frac{3}{2} e_1 nt L \cos(nt + \epsilon - \varpi_1).$$

Here, then, we are met by a difficulty: our equations have been formed on the hypothesis that the square of δr is small enough to be omitted, whereas here, we have a term capable of indefinite increase. This term then, if retained, would ultimately vitiate the whole approximation. The difficulty might, as in the Lunar Theory, be obviated by writing cn for n in the elliptic formulæ, which amounts to supposing the perihelion to be in motion. Its motion is however better found by the method of the variation of elements. Indeed it

---

* These terms are omitted for the sake of brevity: in order, therefore, to obtain the complete second approximation, we must add to the integral of the above equation the result of the first approximation.

DIRECT METHOD OF CALCULATION. 115

may be shewn that such terms as those we are considering, lead to the formulæ which have already been obtained, for the secular variation of the elements*. We shall accordingly altogether neglect such terms, and suppose the elements of the ellipse on which our approximations are based, to have been previously corrected for their secular variations.

113. With this understanding, the complete integral of the equation of Art. 111, will be

$$r\delta r = -\frac{3}{2}\frac{n^2 e_1 P_2}{n^2 - \{(p+1)n - qn'\}^2}$$
$$\cos\left[\{(p+1)n - qn'\}t + Q + \epsilon - \varpi_1\right]$$
$$-\frac{3}{2}\frac{n^2 e_1 P_2}{n^2 - \{(p-1)n - qn'\}}$$
$$\cos\left[\{(p-1)n - qn'\}t + Q - \epsilon + \varpi_1\right]$$
$$+ A\cos(nt - B).$$

If this be added to the result of the first approximation, we obtain for a second approximation

$$r\delta r = \frac{1}{n^2}\left(2m'g + \frac{m'}{2}a\frac{dC_0}{da}\right) + \frac{P_1}{n^2 - (pn - qn')^2}$$
$$\cos\{(pn - qn')t + Q\}$$
$$-\frac{3}{2}\frac{n^2 e_1 P_2}{n^2 - \{(p+1)n - qn'\}^2}$$
$$\cos\left[\{(p+1)n - qn'\}t + Q + \epsilon - \varpi_1\right]$$
$$-\frac{3}{2}\frac{n^2 e_1 P_2}{n^2 - \{(p-1)n - qn'\}^2}$$
$$\cos\left[\{(p-1)n - qn'\}t + Q - \epsilon + \varpi_1\right]$$
$$+ A\cos(nt - B).$$

* It is thus that the Secular Variations are first obtained in the *Mécanique Céleste*. See Pontécoulant's *Système du Monde, Supplément au Livre II.*

8—2

The arbitrary constants might be disposed of as in the first approximation, but it is more convenient in practice to determine them otherwise. In order to obtain a second approximation to the value of δr, it is only necessary to multiply the right-hand member of the above equation by

$$\frac{1}{a}\{1 + e_1 \cos(nt + \epsilon - \varpi_1)\},$$

neglecting e_1^2.

114. *To calculate the perturbations in longitude.*

We have (Art. 106)

$$h\delta\theta = 2\frac{d.r\delta r}{dt} - \delta r\frac{dr}{dt} - 2a\int\frac{dR}{da}dt - 3n\iint\frac{dR}{d\epsilon}dt^2.$$

Taking for simplicity, the first approximation to the value of $r\delta r$, which has been obtained by neglecting the first power of the excentricity, we have

$$r\delta r = \frac{1}{n^2}\left(2m'g + \frac{m'}{2}a\frac{dC_0}{da}\right) + \frac{P_1}{n^2 - (pn - qn')^2}\cos\{(pn - qn')t + Q\};$$

therefore

$$2\frac{d.r\delta r}{dt} = -\frac{2P_1(pn - qn')}{n^2 - (pn - qn')^2}\sin\{(pn - qn')t + Q\}:$$

also, neglecting the first power of the excentricity,

$$\delta r\frac{dr}{dt} = 0.$$

By Art. 108, writing $\frac{1}{2}m'C_0$ for F, we have

$$\frac{dR}{da} = \frac{1}{2}m'\frac{dC_0}{da} + \frac{dP}{da}\cos\{(pn - qn')t + Q\},$$

DIRECT METHOD OF CALCULATION. 117

and $\quad n\int\dfrac{dR}{d\epsilon}dt = \dfrac{nP\dfrac{dQ}{d\epsilon}}{pn-qn'}\cos\{(pn-qn')t+Q\} + m'g;$

therefore $\quad -2a\int\dfrac{dR}{da}dt - 3n\iint\dfrac{dR}{d\epsilon}dt^2$

$= f - \left(m'a\dfrac{dC_0}{da} + 3m'g\right)t$

$\quad -\dfrac{1}{pn-qn'}\left(2a\dfrac{dP}{da} + \dfrac{3nP\dfrac{dQ}{d\epsilon}}{pn-qn'}\right)\sin\{(pn-qn')t+Q\},$

where f is an arbitrary constant.

Hence by substitution, we have

$$h\delta\theta = f - \left(m'a\dfrac{dC_0}{da} + 3m'g\right)t$$

$$-\left\{\dfrac{2a\dfrac{dP}{da}}{pn-qn'} + \dfrac{3nP\dfrac{dQ}{d\epsilon}}{(pn-qn')^2}\right.$$

$$\left. + \dfrac{2P_1(pn-qn')}{n^2-(pn-qn')^2}\right\}\sin\{(pn-qn')t+Q\}.$$

115. This expression is open to the objection of containing a term proportional to the time, which being capable of indefinite increase, would ultimately vitiate the whole approximation. Here, then, we see the advantage of having a quantity g which may be determined at pleasure: we will so determine it that the objectionable term shall vanish. This condition gives

$$g = -\dfrac{1}{3}a\dfrac{dC_0}{da}.$$

We may also omit the constant f, and consider it as contained in the epoch. We have then, writing for h its value $na^2\sqrt{(1-e^2)}$, and neglecting e^2,

$$\delta\theta = -\frac{1}{na^2}\left\{\frac{2a\dfrac{dP}{da}}{pn-qn'} + \frac{3nP\dfrac{dQ}{de}}{(pn-qn')^2}\right.$$
$$\left. + \frac{2P_1(pn-qn')}{n^2-(pn-qn')^2}\right\}\sin\{(pn-qn')t+Q\}.$$

116. Before proceeding to obtain the perturbations in latitude, we will make a few remarks on the forms of the expressions for δr and $\delta\theta$. If we confine ourselves to the results of the first approximation, it will be seen on substituting the value of P_1, that $pn-qn'$ and $n^2-(pn-qn')^2$ occur as divisors, and that the expression for $\delta\theta$ contains besides, the divisor $(pn-qn')^2$. The second of these may be written

$$\{(1-p)n+qn'\}\{(1+p)n-qn'\}.$$

If then either

(i) $pn-qn'$, (ii) $(1-p)n+qn'$, or (iii) $(1-p)n-qn'$,

be very small, the corresponding terms in δr and $\delta\theta$, though of a high order, may yet be sensible. This is especially the case with the first, since as we have remarked, its square occurs in the expression for $\delta\theta$. These are instances of what in the preceding chapter have been characterised as *long inequalities*.

The period of the term $P\cos\{(pn-qn')t+Q\}$, which has given rise to these inequalities is

$$\frac{2\pi}{pn-qn'}:$$

in the case of (i), this is very large, and in that of (ii) or (iii) it is very nearly equal to $\dfrac{2\pi}{n}$, since $pn-qn'$ is nearly equal to $\pm n$. Hence it appears that terms in R whose period is either very large, or nearly equal to that of the planet, may give rise to important inequalities in the radius vector and longitude. Their actual importance will of course depend in part upon the order of the principal part of P_1 with respect to the excentricities and inclinations, i.e. (see Art. 50) upon $p \sim q$.

117. *To integrate the equation for the perturbation in latitude.*

The equation is (Art. 107)
$$\frac{d^2(r\delta\lambda)}{dt^2} + \frac{\mu}{r^3}(r\delta\lambda) = \frac{dR}{dz},$$
the position of the plane of the orbit of the disturbed planet at the epoch, being taken for the fixed plane of reference.

Differentiating the expression for R in Art. 44, with respect to z, we obtain
$$-m'(z-z')(\tfrac{1}{2}D_0 + D_1 \cos\phi + \ldots + D_k \cos k\phi + \ldots) - \frac{m'z'}{a^3},$$
or, putting z equal to 0, and substituting
$$a' \tan i'' \sin(n't + \epsilon' - \Omega')$$
for z' (see Art. 42),
$$-m'a' \tan i'' \sin(n't + \epsilon' - \Omega') \left\{\frac{1}{2}D_0 + \frac{1}{a'^3} + D_1 \cos\phi + \ldots\right\}.$$

This expression, after reduction, consists of terms of the form
$$P \sin\{(pn - qn')t + Q\},$$
where p and q are positive integers, and either may be zero. Considering one such term, our equation becomes
$$\frac{d^2(r\delta\lambda)}{dt^2} + \frac{\mu}{r^3}(r\delta\lambda) = P\sin\{(pn - qn')t + Q\}.$$

Now as in Art. 108,

$$\frac{\mu}{r^3} = n^2 \{1 + 3e \cos(nt + \epsilon - \varpi) + \ldots\};$$

hence, neglecting the product $e\delta\lambda$,

$$\frac{d^2(r\delta\lambda)}{dt^2} + n^2 . r\delta\lambda = P \sin\{(pn - qn')t + Q\}.$$

The integral of this equation is

$$r\delta\lambda = \frac{P}{n^2 - (pn - qn')^2} \sin\{(pn - qn')t + Q\} + A\cos(nt - B).$$

If instead of taking for the fixed plane of reference, the plane of the orbit of the disturbed planet at epoch, we take a plane slightly inclined to this, we may omit the arbitrary term. For, denoting the planet's latitude with respect to this plane by λ, we have approximately

$$\lambda = \tan i \sin(nt + \epsilon - \Omega),$$

and it may be shewn as in Art. 110, that omitting the term in question is only equivalent to changing slightly the values of i and Ω.

CHAPTER VIII.

ON THE EFFECTS WHICH A RESISTING MEDIUM WOULD PRODUCE IN THE MOTIONS OF THE PLANETS.

118. In the preceding chapters, we have supposed the planetary motions to take place in free space, and the results of calculations based upon this hypothesis manifest a very close agreement with observation. There is, however, a remarkable circumstance connected with Encke's comet which seems to indicate the possibility of the existence of a very rare medium, too rare indeed to cause any sensible resistance to the motions of the planets, but which, as we shall presently see, may yet influence the motions of comets, in consequence of the extreme smallness of the masses of these bodies. It has been observed that the comet above referred to (which describes an elliptic orbit in a period of about $3\frac{1}{4}$ years,) has since its appearance in 1786, been moving round the Sun with an increasing mean motion. Encke attributes this to the resistance of a medium pervading space. We shall therefore proceed to examine the effects which such a medium would produce upon the elements of a planet's orbit, assuming the resistance to vary as the product of the density of the medium and the square of the velocity of the planet. We shall neglect, in the present investigation, all forces except the Sun's attraction and the resistance of the medium; consequently the planet may be supposed to move wholly in one plane.

119. Let r, θ be the radius vector and longitude of the planet, s the length of an arc of its actual orbit measured from some fixed point to its position at time t, and ρ the density of the medium. Then if k be a constant, we may represent the resistance on the planet by $k\rho \left(\dfrac{ds}{dt}\right)^2$, and the equations of motion will be

$$\frac{d^2r}{dt^2} - r\left(\frac{d\theta}{dt}\right)^2 = -\frac{\mu}{r^2} - k\rho\left(\frac{ds}{dt}\right)^2 \frac{dr}{ds},$$

$$\frac{1}{r}\frac{d}{dt}\left(r^2\frac{d\theta}{dt}\right) = -k\rho\left(\frac{ds}{dt}\right)^2 r\frac{d\theta}{ds}.$$

If $r^2\dfrac{d\theta}{dt} = h$, these may be written

$$\frac{d^2r}{dt^2} - r\left(\frac{d\theta}{dt}\right)^2 = -\frac{\mu}{r^2} - k\rho\frac{ds}{dt}\frac{dr}{dt} \quad \ldots\ldots\ldots\ldots (1),$$

$$\frac{d}{dt}\left(r^2\frac{d\theta}{dt}\right) = -k\rho h\frac{ds}{dt} \quad \ldots\ldots\ldots\ldots\ldots\ldots (2).$$

These equations are the same in form with those of Art. 20, and may be treated in a similar manner, $-k\rho\dfrac{ds}{dt}\dfrac{dr}{dt}$ taking the place of $\dfrac{dR}{dr}$, and $-k\rho h\dfrac{ds}{dt}$ that of $\dfrac{dR}{d\theta}$. We have from equation (2)

$$\frac{dh}{dt} = -k\rho h\frac{ds}{dt}.$$

120. *To obtain a formula for calculating the mean distance.*

We might proceed as in Art. 25, but we shall here employ the method of Art. 26. We have

$$\frac{d^2s}{dt^2} = -\frac{\mu}{r^2}\frac{dr}{ds} - k\rho\left(\frac{ds}{dt}\right)^2,$$

and by a known formula of elliptic motion

$$\left(\frac{ds}{dt}\right)^2 = \frac{2\mu}{r} - \frac{\mu}{a}.$$

Differentiating the latter, we obtain

$$2\frac{ds}{dt}\frac{d^2s}{dt^2} = -\frac{2\mu}{r^2}\frac{dr}{dt} + \frac{\mu}{a^2}\frac{da}{dt},$$

and multiplying the former by $2\dfrac{ds}{dt}$,

$$2\frac{ds}{dt}\frac{d^2s}{dt^2} = -\frac{2\mu}{r^2}\frac{dr}{dt} - 2k\rho\left(\frac{ds}{dt}\right)^2;$$

therefore
$$\frac{\mu}{a^2}\frac{da}{dt} = -2k\rho\left(\frac{ds}{dt}\right)^2,$$

or
$$\frac{da}{dt} = -\frac{2k\rho a^2}{\mu}\left(\frac{ds}{dt}\right)^3.$$

121. *To obtain a formula for calculating the excentricity.*

We have, as in Art. 27,

$$\frac{\mu^2 e^2}{h^2} = \left(\frac{dr}{dt}\right)^2 + \left(\frac{h}{r} - \frac{\mu}{h}\right)^2 \dots\dots\dots\dots (3).$$

Differentiating as if r were constant, and writing $-k\dfrac{dr}{dt}\dfrac{ds}{dt}$ for $\dfrac{d^2r}{dt^2}$,

$$\frac{\mu^2 e}{h^2}\frac{de}{dt} = -k\rho\left(\frac{dr}{dt}\right)^2\frac{ds}{dt} + \left\{\left(\frac{h}{r} - \frac{\mu}{h}\right)\left(\frac{1}{r} + \frac{\mu}{h^2}\right) + \frac{\mu^2 e^2}{h^3}\right\}\frac{dh}{dt}$$

$$= -k\rho\left(\frac{dr}{dt}\right)^2\frac{ds}{dt} - \left\{\frac{h}{r^2} - \frac{\mu^2(1-e^2)}{h^3}\right\}k\rho h\frac{ds}{dt}$$

$$= -k\rho\frac{ds}{dt}\left\{\left(\frac{dr}{dt}\right)^2 + \frac{h^2}{r^2} - \frac{\mu^2(1-e^2)}{h^2}\right\}.$$

Now from equation (3)

$$\left(\frac{dr}{dt}\right)^2 + \frac{h^2}{r^2} = \frac{\mu^2 e^2}{h^2} + \frac{2\mu}{r} - \frac{\mu^2}{h^2}$$

$$= \frac{2\mu}{r} - \frac{\mu^2(1-e^2)}{h^2};$$

therefore
$$\frac{\mu^2 e}{h^2}\frac{de}{dt} = -2k\rho\frac{ds}{dt}\left\{\frac{\mu}{r} - \frac{\mu^2(1-e^2)}{h^2}\right\},$$

or
$$\frac{de}{dt} = -\frac{2k\rho}{e}\frac{ds}{dt}\left\{\frac{h^2}{\mu r} - (1-e^2)\right\}$$

$$= -\frac{2k\rho(1-e^2)}{e}\left(\frac{a}{r} - 1\right)\frac{ds}{dt}.$$

This result may also be obtained by differentiating the formula $h^2 = \mu a(1-e^2)$, and substituting the expressions for $\frac{dh}{dt}$ and $\frac{da}{dt}$, as in Art. 28.

122. *To obtain a formula for calculating the longitude of perihelion.*

We have, as in Art. 29,

$$\frac{dr}{dt}\cot(\theta - \varpi) = \frac{h}{r} - \frac{\mu}{h} \quad \text{.................... (4).}$$

Differentiating as if r and θ were constant, and writing $-k\rho\frac{dr}{dt}\frac{ds}{dt}$ for $\frac{d^2r}{dt^2}$,

$$\frac{dr}{dt}\operatorname{cosec}^2(\theta - \varpi)\frac{d\varpi}{dt} - k\rho\frac{dr}{dt}\frac{ds}{dt}\cot(\theta - \varpi) = \left(\frac{1}{r} + \frac{\mu}{h^2}\right)\frac{dh}{dt},$$

or since
$$\frac{dh}{dt} = -k\rho h\frac{ds}{dt},$$

$$\frac{dr}{dt} \operatorname{cosec}^2 (\theta - \varpi) \frac{d\varpi}{dt} = k\rho \frac{ds}{dt} \left\{ \frac{dr}{dt} \cot (\theta - \varpi) - \left(\frac{h}{r} + \frac{\mu}{h}\right) \right\}$$

$$= -2 \frac{k\rho\mu}{h} \frac{ds}{dt}, \text{ by (4)};$$

but from equation (4) of Art. 22,

$$\frac{dr}{dt} \operatorname{cosec} (\theta - \varpi) = \frac{\mu e}{h};$$

therefore
$$\frac{d\varpi}{dt} = -\frac{2k\rho}{e} \sin (\theta - \varpi) \frac{ds}{dt}.$$

123. Before proceeding to obtain a formula for calculating the epoch, we shall express the results of the preceding articles in a form convenient for application.

If u denote the excentric anomaly, we have

$$r = a(1 - e \cos u) \dots\dots\dots\dots\dots\dots (1),$$

$$nt + \epsilon - \varpi = u - e \sin u \dots\dots\dots\dots\dots (2).$$

Hence
$$\left(\frac{ds}{dt}\right)^2 = \frac{2\mu}{r} - \frac{\mu}{a}$$

$$= \frac{\mu}{a} \left(\frac{2}{1 - e \cos u} - 1\right)$$

$$= \frac{\mu}{a} \frac{1 + e \cos u}{1 - e \cos u};$$

therefore
$$\frac{ds}{dt} = na \sqrt{\left(\frac{1 + e \cos u}{1 - e \cos u}\right)}, \text{ since } na^{\frac{3}{2}} = \sqrt{\mu}.$$

From (1), by differentiation as if the elements were invariable, we have

$$\frac{dr}{dt} = ae \sin u \frac{du}{dt};$$

and from the equation
$$\frac{1}{r} = \frac{\mu}{h^2}\{1 + e\cos(\theta - \varpi)\},$$
$$\frac{dr}{dt} = \frac{\mu e}{h}\sin(\theta - \varpi);$$

equating the two values of $\frac{dr}{dt}$,
$$\sin(\theta - \varpi) = \frac{ha}{\mu}\sin u \frac{du}{dt}.$$

From (2), by differentiation as if the elements were invariable, we have
$$\frac{dt}{du} = \frac{1}{n}(1 - e\cos u).$$

Hence, by substitution, the formulæ of the preceding articles become
$$\frac{da}{du} = -2k\rho a^2 (1 + e\cos u)\sqrt{\left(\frac{1 + e\cos u}{1 - e\cos u}\right)},$$
$$\frac{de}{du} = -2k\rho a(1 - e^2)\cos u \sqrt{\left(\frac{1 + e\cos u}{1 - e\cos u}\right)},$$
$$\frac{d\varpi}{du} = -\frac{2k\rho a\sqrt{(1 - e^2)}}{e}\sin u\sqrt{\left(\frac{1 + e\cos u}{1 - e\cos u}\right)}.$$

124. *To obtain a formula for calculating the epoch.*

By differentiating the formulæ
$$r = a(1 - e\cos u),$$
$$nt + \epsilon - \varpi = u - e\sin u,$$
considering the elements, first variable, and then invariable, and comparing the results, we obtain
$$ae\sin u \frac{du}{dt} = a\cos u \frac{de}{dt} - (1 - e\cos u)\frac{da}{dt},$$

$$t\frac{dn}{dt} + \frac{d\epsilon}{dt} - \frac{d\varpi}{dt} = (1 - e\cos u)\frac{du}{dt} - \sin u \frac{de}{dt};$$

whence, eliminating $\dfrac{du}{dt}$

$$t\frac{dn}{dt} + \frac{d\epsilon}{dt} - \frac{d\varpi}{dt} = \frac{\cos u - e}{e \sin u}\frac{de}{dt} - \frac{(1 - e\cos u)^2}{ae \sin u}\frac{da}{dt}.$$

As in Art. 37, we may omit the term $t\dfrac{dn}{dt}$ if we bear in mind that the mean longitude will then be denoted by $\int n dt + \epsilon$. Thus, on substituting the values of $\dfrac{da}{dt}$, $\dfrac{de}{dt}$, and $\dfrac{d\varpi}{dt}$, and reducing, we obtain

$$\frac{d\epsilon}{du} = \frac{2k\rho a}{e}\left\{1 - \sqrt{(1 - e^2)} - e^3 \cos u\right\} \sin u \sqrt{\left(\frac{1 + e\cos u}{1 - e\cos u}\right)}.$$

125. The formulæ of the preceding articles are sufficient to determine the elements of the orbit at any time, and being perfectly general, are applicable as well to the motion of comets, as to that of planets, but before we can integrate them, we shall require a knowledge of the form of ρ. Now the analogy of the terrestrial atmosphere would lead us to suppose that if the sun be surrounded by an ethereal medium, its density decreases as the distance from the sun increases. Moreover, the researches of Professor Encke on the comet which bears his name, seem to indicate the law of the inverse square. We will, however, assume ρ to be such a function of r, that when multiplied by $\sqrt{\left(\dfrac{1 + e\cos u}{1 - e\cos u}\right)}$, and developed in a series of cosines of u and its multiples, it takes the form

$$A + Be\cos u + Ce^2 \cos 2u + \ldots$$

Thus our formulæ become

$$\frac{da}{du} = -2ka^2\{A + (A+B)e\cos u + \ldots\},$$

$$\frac{de}{du} = -2ka\left\{A\cos u + \frac{Be}{2}(1+\cos 2u) + \ldots\right\},$$

$$e\frac{d\varpi}{du} = -2ka\left\{A\sin u + \frac{Be}{2}\sin 2u + \ldots\right\},$$

$$\frac{d\epsilon}{du} = ka(Ae\sin u + \ldots).$$

126. *Supposing the orbit nearly circular, to examine the effect of the medium upon the elements of the orbit.*

Since the orbit is nearly circular, we shall neglect squares and higher powers of e; thus the preceding formulæ give on integration

$$a = \text{const.} - 2ka^2\{Au + (A+B)e\sin u\},$$

$$e = \text{const.} - 2ka\left\{A\sin u + \frac{Be}{2}\left(u + \frac{\sin 2u}{2}\right)\right\},$$

$$e\varpi = \text{const.} + 2ka\left\{A\cos u + \frac{Be}{4}\cos 2u\right\},$$

$$\epsilon = \text{const.} - ka\,Ae\cos u.$$

Hence in an entire revolution of the planet, the mean distance is diminished by $4\pi ka^2 A$, and the excentricity by $2\pi kaBe$, while the longitudes of perihelion, and of the epoch remain unchanged. Also from the formula $n = \dfrac{\sqrt{\mu}}{a^{\frac{3}{2}}}$, it appears that the mean motion is, in an entire revolution, increased by $6\pi knaA$.

127. We have already remarked that no traces of a resisting medium have yet been discovered in the motion of the planets: but, since k varies inversely as the mass of the body acted upon, the formulæ of Art. 123 shew that such a medium, though too rare to influence the planets, might yet sensibly affect the motions of comets, in consequence of the extreme smallness of their masses.

PROBLEMS.

1. SUPPOSING in the Problem of the Three Bodies the relative orbit of two of the bodies to be a circle described uniformly, obtain equations for determining the motion of the third body; and transform the system of co-ordinates, so that the plane of the circular orbit being that of xy, the axis of x shall always pass through the two bodies in that plane.

2. Shew that the plane of the orbit of a planet revolves about the planet's radius vector as an instantaneous axis*.

3. A particle is describing an orbit round a centre of force which is any function of the distance, and is acted upon by a disturbing force which is always perpendicular to the plane of the instantaneous orbit, and inversely proportional to the distance of the body from the centre of the principal force. Prove that the plane of the instantaneous orbit revolves uniformly round its instantaneous axis.

4. Find when the curvature of the instantaneous orbit of a body, acted on by disturbing forces, is the same as that of the actual orbit; and shew that this is always the case when the only disturbing force arises from the action of a resisting medium.

---

* In this and the following problem, the plane of the orbit must be supposed to have no angular velocity about a normal to itself. See note to Art. 19.

PROBLEMS.

5. If R be expressed on the one hand as a function of r_1, θ_1, and z (Art. 11), and on the other as a function of r, θ, i, and Ω, θ being measured on the plane of the orbit *from the node*, prove that

$$\frac{dR}{d\theta_1} = \frac{dR}{d\Omega},$$

and obtain a formula for calculating the inclination.

6. If R be expressed as a function of t and the usual elements, obtain the formulæ

$$\frac{d\Omega}{dt} = \frac{1}{h \sin i} \frac{dR}{di},$$

$$\frac{di}{dt} = \frac{\cot(\theta - \Omega)}{h} \frac{dR}{di},$$

where θ is measured on the plane of reference as far as the node, and thence on that of the orbit, and

$$h = r^2 \left(\frac{d\theta}{dt} - 2 \sin^2 \frac{i}{2} \frac{d\Omega}{dt}\right).$$

7. The central force being $\frac{\mu}{r^2} + \frac{\mu'}{r^1}$, obtain the following equation for the apsidal motion

$$\frac{d\varpi}{dt} = \frac{\sqrt{a(1-e^2)}}{e\sqrt{\mu}} \frac{\mu' \cos(\theta - \varpi)}{r^2},$$

a, e and ϖ being elements of the instantaneous ellipse.

8. A body revolving about a centre of force, which varies inversely as the square of the distance, is constantly retarded by a small constant force; find the alteration of the major axis, excentricity, and apse, in one revolution.

9. When the disturbing function R is independent of θ, find expressions for $\dfrac{de}{dt}$ and $\dfrac{d\varpi}{dt}$.

If $R = \dfrac{m'}{r}$, these expressions give variable values for e and ϖ, whereas the motion of the body actually takes place in a fixed ellipse: shew this, and explain the apparent paradox.

10. A planet describes an orbit under the action of a force $\dfrac{\mu}{r^2}$ tending to the Sun, μ not being quite constant: obtain the following equations for the variations of the excentricity and longitude of perihelion;

$$\frac{d(\mu e)}{d\mu} = -\cos(\theta - \varpi),$$

$$\mu e \frac{d\varpi}{d\mu} = -\sin(\theta - \varpi).$$

If $d\mu$ be always positive, what in a whole revolution is the nature of its effects upon the excentricity and position of the major axis?

11. If the equation of the Moon's orbit be reduced to the form

$$\frac{d^2u}{d\theta^2} + u - a = af,$$

shew that the excentricity and longitude of perihelion may be found from the equations

$$\frac{de}{d\theta} = -f \sin(\theta - \varpi), \quad e \frac{d\varpi}{d\theta} = f \cos(\theta - \varpi).$$

Apply these equations to find e and ϖ, when f is a small disturbing force, depending only upon the Moon's distance from the Earth.

PROBLEMS. 133

12. Assuming the differential equation for s in the Lunar Theory to be

$$\frac{d^2s}{d\theta^2} + s = -m^2 s \left\{ \frac{3}{2} + \frac{3}{2} \cos 2(\theta - m\theta) \right\}$$
$$+ m^2 \frac{ds}{d\theta} \left\{ \frac{3}{2} \sin 2(\theta - m\theta) \right\},$$

shew that if γ be the longitude of the Moon's node,

$$\frac{d\gamma}{d\theta} = -\frac{3}{4} m^2 \{1 - \cos 2(m\theta - \gamma) - \cos 2(\theta - \gamma)$$
$$+ \cos 2(\theta - m\theta)\}.$$

From the above expression for $\frac{d\gamma}{d\theta}$, find the ratio of the mean motion of the node to that of the Moon, taking into account terms of the order m^4.

13. If two planets disturbing one another were revolving in periods of 350 and 201 days, what form of terms in the disturbing function would demand examination?

14. The periods of Venus and the Earth are 224·7 and 365·256 days respectively; find approximately the period of the long inequality arising from their mutual perturbations, the important term in the disturbing function R being of the form

$$Pe^3 e'^2 \cos \{13(nt + \epsilon) - 8(n't + \epsilon') - 3\varpi - 2\varpi'\}.$$

15. The radius vector of a planet is affected with a small periodical inequality; shew that its effect may be represented by continued and periodical alterations of the excentricity and longitude of perihelion, the period of either being $\frac{PT}{P \sim T}$, where P is the period of the planet and T that of the inequality.

16. If in addition to the force of the Sun on a planet there be a small force tending towards the Sun, and varying inversely as the mth power of the distance of the planet from the Sun, prove that the perihelion of the orbit will have a progressive or regressive motion according as m is greater or less than 2.

Can you explain this result by reasoning similar to that used in Airy's *Gravitation?*

17. It has been found by comparing theory with observation that the perihelion of Mercury progresses at a rate greater by a than that due to the attraction of known bodies: shew that this increment would be accounted for if the law of force tending to the Sun were $\dfrac{\mu}{r^2} + \dfrac{\mu'}{r^4}$, and if $\mu' = ac^4 \sqrt{\dfrac{\mu}{c}}$, the orbit being supposed to be nearly a circle, and the mean distance to be c.

18. The central force acting on a body being

$$\frac{\mu}{r^2} + \mu' \phi(r)$$

shew to terms inclusive of μ' and the square of the excentricity, that the motion is in an ellipse revolving uniformly about the focus.

19. Shew by means of the formula

$$\frac{da}{dt} = \frac{2na^2}{\mu} \frac{dR}{d\epsilon}$$

that the chief perturbation of the axis major of the Moon's orbit may be expressed by the equation

$$a_1 = a \left\{ 1 + \frac{3n'^2}{2n(n-n')} \cos 2(nt + \epsilon - n't - \epsilon') \right\},$$

where n and n' are the mean motions of the Moon and Sun respectively.

20. A satellite revolving in an ellipse of small excentricity is disturbed by another satellite revolving about the same primary; find approximately the variation of the mean distance and the motion of the apse, corresponding to the terms

$$\frac{n'^2}{4} r^2 [1 + 3 \cos \{2(n-n')t + \epsilon - \epsilon'\}]$$

in the function R, having given

$$\frac{da}{dt} = \frac{2na^2}{\mu} \frac{dR}{d\epsilon}, \quad \frac{d\varpi}{dt} = \frac{na\sqrt{(1-e^2)}}{\mu e} \frac{dR}{de}.$$

21. Prove that, neglecting periodical variations, the excentricity of any orbit can always be represented by the diagonal of a parallelogram, whose sides are constant, and angle varies uniformly.

22. Given the equations

$$\tan^2 i = N_1^2 + N_2^2 + 2N_1 N_2 \cos(h_1 t + \delta_1 - \delta_2),$$

$$\tan \Omega = \frac{N_1 \sin(h_1 t + \delta_1) + N_2 \sin \delta_2}{N_1 \cos(h_1 t + \delta_1) + N_2 \cos \delta_2};$$

explain the nature of the motion of the node, when the minimum inclination is zero.

23. Prove that as far as secular variations only are concerned the function F is constant.

24. Considering only secular variations, obtain the following equations:

$$\Sigma \left(\frac{m}{na} e^2 \frac{d\varpi}{dt} \right) = C, \quad \Sigma \left(\frac{m}{na} \tan^2 i \frac{d\Omega}{dt} \right) = C.$$

25. If the squares of the masses of two mutually disturbing planets were to each other inversely as their mean distances, shew that the nodes would oscillate through equal angles.

26. If M, m, m' be the masses of three bodies mutually attracting according to the law of gravity, M being much larger than m or m', and if v, v' be the velocities of m, m' at distances r, r' from the centre of M, supposed fixed, shew that the equation of *vis viva* for this case may be assumed to be

$$mv^2 + m'v'^2 + 2M\left(\frac{m}{2a} - \frac{m}{r} + \frac{m'}{2a'} - \frac{m'}{r'}\right) = 0,$$

$2a$ and $2a'$ being the major axes of the instantaneous ellipses of m and m'.

27. Infer from the foregoing equation by the method of the variation of parameters the ratio of simultaneous changes in the mean distances and mean motions of two planets mutually disturbing.

28. If r be the true radius vector, θ_1 the projected longitude, and λ the latitude of a planet, obtain the following equation of motion:

$$\frac{d^2r}{dt^2} - r\cos^2\lambda \left(\frac{d\theta_1}{dt}\right)^2 - r\left(\frac{d\lambda}{dt}\right)^2 + \frac{\mu}{r^2} = \frac{dR}{dr}.$$

29. Obtain the following equation between the perturbations of a planet in longitude and radius vector, whatever be the law of force, provided it be central and a function of the distance only, and provided such a function as R can be found:

$$h \cdot \delta\theta = \frac{d}{dt}(2r\,\delta r) - \frac{1}{r}\frac{dr}{dt}r\,\delta r + 3\int \frac{d(R)}{dt}\,dt + 2r\frac{dR}{dr}$$
$$- 4F\delta r - 2r\frac{dF}{dr}\delta r,$$

where F denotes the central force, and h twice the sectorial area described by the undisturbed planet round the Sun.

30. If the orbits of two planets which disturb each other be very nearly circular, shew that the inequalities of the radius vector may be immediately deduced from those of the longitude by means of the equation

$$\frac{\delta r}{r} + \frac{1}{2}\frac{\dfrac{d.\delta\theta}{dt}}{\dfrac{d\theta}{dt}} - \frac{1}{2}\frac{a}{\mu}\frac{n}{n-n'}R = 0.$$

31. Integrate the equation

$$\frac{d^2(r\delta r)}{dt^2} + n^2 . r\delta r = \Sigma\{P\cos(pnt + Q)\},$$

determining the arbitrary constants so that $\delta r = 0$, and $\dfrac{d.\delta r}{dt} = 0$, when $t = 0$: and shew that for small values of t,

$$r\delta r = \Sigma\left(\frac{Pt^2}{2}\cos Q\right),$$

the case of $p = 1$ being included.

32. A planet moves in a resisting medium of which the resistance

$$= \frac{f}{r^2}\left(\frac{ds}{dt}\right)^2;$$

apply the equation

$$\frac{d^2(r\delta r)}{dt^2} + \frac{\mu}{r^3}. r\delta r - 2\int\frac{d(R)}{dt}dt - r\frac{dR}{dr} = 0$$

to obtain the following, in which e^2 is neglected:

$$\frac{d^2(r\delta r)}{dt^2} + n^2. r\delta r + n^2. r\delta r . 3e\cos(nt + \epsilon - \varpi)$$

$$+ 2fn^2a\left\{nt + \frac{11}{2}e\sin(nt + \epsilon - \varpi)\right\} = 0.$$

33. The co-ordinates of the position at any time t of a disturbed planet being $x+\delta x$, $y+\delta y$, $z+\delta z$, reckoned from the Sun's centre as a fixed origin, and referred to the plane of motion at a given epoch; and r being the heliocentric distance, x, y the co-ordinates of the position which the planet would have had at the time t, if the disturbance had ceased at the given epoch; obtain the following equations for determining δx, δy, δz to the first order of the disturbing force:

$$\frac{d^2(\delta x)}{dt^2} + \frac{\mu}{r^3}\left\{\left(1-\frac{3x^2}{r^2}\right)\delta x - \frac{3xy}{r^2}\delta y\right\} + \frac{dR'}{dx'} = 0,$$

$$\frac{d^2(\delta y)}{dt^2} + \frac{\mu}{r^3}\left\{\left(1-\frac{3y^2}{r^2}\right)\delta y - \frac{3xy}{r^2}\delta x\right\} + \frac{dR'}{dy'} = 0,$$

$$\frac{d^2(\delta z)}{dt^2} + \frac{\mu \delta z}{r^3} + \frac{dR'}{dz'} = 0,$$

in which μ is the sum of the masses of the Sun and planet, and R' is put for

$$-m'(x'^2+y'^2+z'^2)^{-\frac{1}{2}} + m'\{(x-x')^2+(y-y')^2+(z-z')^2\}^{-\frac{1}{2}},$$

m' being the mass, and x', y', z' the heliocentric co-ordinates of the disturbing planet.

34. Shew that the effect of a resisting medium on the instantaneous orbit of a planet, would be to make the apsidal line regrede or progrede, according as the planet moved from perihelion to aphelion, or from aphelion to perihelion.

35. Two small planets P, Q, very near each other, revolve about the Sun in orbits very nearly circular, and make two revolutions about each other while they make one revolution about the Sun. Compare the sum of their masses with the mass of the Sun.

* If the line PQ move parallel to itself, what inference do you draw?

36. If the motion of a planet round the Sun be disturbed by the action of another planet, the latter being supposed to describe a circular orbit of radius a' with uniform velocity n, obtain the following exact equation:

$$\left(\frac{dr_1}{dt}\right)^2 + r_1^2\left(\frac{d\theta_1}{dt}\right)^2 + \left(\frac{dz}{dt}\right)^2 - 2nr_1^2\frac{d\theta_1}{dt} + C$$
$$= \frac{2\mu}{r} - \frac{2m'r}{a'^2}\cos\omega + \frac{2m'}{(a'^2 - 2a'r\cos\omega + r^2)^{\frac{1}{2}}},$$

where r is the radius vector of the disturbed planet, r_1, θ_1, z its co-ordinates referred to a fixed plane, and ω the inclination of the radii vectores of the disturbed and disturbing planets to each other.

37. Prove that, if the periodic times of a disturbed and disturbing planet are not commensurable, the secular changes of the orbit of the disturbed planet are the same as they would be if the mass of the disturbing planet were distributed over its orbit, in such a manner, that the part of the mass distributed over each portion of the orbit should be proportional to the time which the planet actually takes to describe that portion.

APPENDIX.

ON THE FORM OF THE EQUATIONS OF ART. 39.

1. On referring to Art. 39, it will be seen that the formulæ which have been obtained for calculating the elements of the orbit involve only partial differential coefficients of R with respect to these elements, multiplied by coefficients which do not contain the time explicitly. This circumstance greatly facilitates their application, by rendering them fit for use as soon as the partial differential coefficients have been calculated. We proceed to shew that this remarkable characteristic is not restricted to the particular system of elements which have been adopted, but may be attained with any system of elements whatever. The discovery of this is due to Lagrange.

2. If the motion of the planet be referred to three rectangular axes originating in the centre of gravity of the Sun, we have the equations of motion (see Art. 9)

$$\frac{d^2x}{dt^2} + \frac{\mu x}{r^3} = \frac{dR}{dx} \quad \ldots\ldots\ldots\ldots\ldots\ldots (1),$$

$$\frac{d^2y}{dt^2} + \frac{\mu y}{r^3} = \frac{dR}{dy} \quad \ldots\ldots\ldots\ldots\ldots\ldots (2),$$

$$\frac{d^2z}{dt^2} + \frac{\mu z}{r^3} = \frac{dR}{dz} \quad \ldots\ldots\ldots\ldots\ldots\ldots (3).$$

APPENDIX. 141

Let a, b, c, d, e, f be the six elements introduced by integrating these equations when $R = 0$, and for $\dfrac{dx}{dt}, \dfrac{dy}{dt}, \dfrac{dz}{dt}$ write x', y', z': then x', y', and z' can be expressed as functions of t and the elements; hence

$$\frac{dx'}{dt} = \left(\frac{dx'}{dt}\right) + \frac{dx'}{da}\frac{da}{dt} + \frac{dx'}{db}\frac{db}{dt} + \ldots\ldots,$$

where in $\left(\dfrac{dx'}{dt}\right)$ the elements are supposed constant.

If in equation (1) we put R equal to 0, we have

$$\left(\frac{dx'}{dt}\right) + \frac{\mu x}{r^3} = 0 ;$$

therefore
$$\frac{dx'}{dt} - \left(\frac{dx'}{dt}\right) = \frac{dR}{dx} ;$$

therefore
$$\frac{dR}{dx} = \frac{dx'}{da}\frac{da}{dt} + \frac{dx'}{db}\frac{db}{dt} + \ldots\ldots,$$

and similar equations hold for $\dfrac{dR}{dy}$ and $\dfrac{dR}{dz}$.

Now since R is a function of x, y, and z,

$$\frac{dR}{da} = \frac{dR}{dx}\frac{dx}{da} + \frac{dR}{dy}\frac{dy}{da} + \frac{dR}{dz}\frac{dz}{da}$$

$$= \left(\frac{dx}{da}\frac{dx'}{da} + \frac{dy}{da}\frac{dy'}{da} + \frac{dz}{da}\frac{dz'}{da}\right)\frac{da}{dt}$$

$$+ \left(\frac{dx}{da}\frac{dx'}{db} + \frac{dy}{da}\frac{dy'}{db} + \frac{dz}{da}\frac{dz'}{db}\right)\frac{db}{dt}$$

$$+ \left(\frac{dx}{da}\frac{dx'}{dc} + \frac{dy}{da}\frac{dy'}{dc} + \frac{dz}{da}\frac{dz'}{dc}\right)\frac{dc}{dt}$$

$$+ \ldots\ldots\ldots\ldots\ldots\ldots\ldots\ldots\ldots$$

3. We may eliminate $\dfrac{da}{dt}$ from this expression: for, supposing x, y, and z expressed as functions of t and the elements, we have
$$\frac{dx}{dt} = \left(\frac{dx}{dt}\right) + \frac{dx}{da}\frac{da}{dt} + \frac{dx}{db}\frac{db}{dt} + \ldots;$$
but by the principles of the method of the Variation of Parameters
$$\frac{dx}{dt} = \left(\frac{dx}{dt}\right);$$
therefore $\quad \dfrac{dx}{da}\dfrac{da}{dt} + \dfrac{dx}{db}\dfrac{db}{dt} + \dfrac{dx}{dc}\dfrac{dc}{dt} + \ldots = 0.$

Similarly, $\dfrac{dy}{da}\dfrac{da}{dt} + \dfrac{dy}{db}\dfrac{db}{dt} + \dfrac{dy}{dc}\dfrac{dc}{dt} + \ldots = 0,$

$\qquad\quad \dfrac{dz}{da}\dfrac{da}{dt} + \dfrac{dz}{db}\dfrac{db}{dt} + \dfrac{dz}{dc}\dfrac{dc}{dt} + \ldots = 0.$

Multiplying these equations by $\dfrac{dx'}{da}$, $\dfrac{dy'}{da}$, &c., and adding, we obtain
$$0 = \left(\frac{dx}{da}\frac{dx'}{da} + \frac{dy}{da}\frac{dy'}{da} + \frac{dz}{da}\frac{dz'}{da}\right)\frac{da}{dt}$$
$$+ \left(\frac{dx}{db}\frac{dx'}{da} + \frac{dy}{db}\frac{dy'}{da} + \frac{dz}{db}\frac{dz'}{da}\right)\frac{db}{dt}$$
$$+ \left(\frac{dx}{dc}\frac{dx'}{da} + \frac{dy}{dc}\frac{dy'}{da} + \frac{dz}{dc}\frac{dz'}{da}\right)\frac{dc}{dt}$$
$$+ \ldots\ldots\ldots\ldots$$

If this expression be subtracted from that for $\dfrac{dR}{da}$ in Art. 2, the latter may be written
$$\frac{dR}{da} = [a,\,b]\frac{db}{dt} + [a,\,c]\frac{dc}{dt} + \ldots$$

where $[a, b] = \dfrac{dx}{da}\dfrac{dx'}{db} - \dfrac{dx}{db}\dfrac{dx'}{da} + \dfrac{dy}{da}\dfrac{dy'}{db} - \dfrac{dy}{db}\dfrac{dy'}{da}$

$$+ \dfrac{dz}{da}\dfrac{dz'}{db} - \dfrac{dz}{db}\dfrac{dz'}{da}.$$

Similarly,
$$\dfrac{dR}{db} = [b, a]\dfrac{da}{dt} + [b, c]\dfrac{dc}{dt} + \ldots$$

4. By successive elimination between these equations, we can obtain expressions for $\dfrac{da}{dt}$, $\dfrac{db}{dt}$, &c., in terms of $\dfrac{dR}{da}$, $\dfrac{dR}{db}$, &c., $[a, b]$, $[a, c]$, &c.: if, then, we can shew that $[a, b]$, $[a, c]$, &c., are independent of the time explicitly, it will follow that this is also the case with the coefficients of $\dfrac{dR}{da}$, $\dfrac{dR}{db}$, &c., in the expressions for $\dfrac{da}{dt}$, $\dfrac{db}{dt}$, &c.

5. *To shew that* [a, b] *is independent of the time explicitly.*

Let $V = \dfrac{\mu}{r}$; then the equations of motion give

$$\left(\dfrac{dx'}{dt}\right) = \dfrac{dV}{dx}, \quad \left(\dfrac{dy'}{dt}\right) = \dfrac{dV}{dy}, \quad \left(\dfrac{dz'}{dt}\right) = \dfrac{dV}{dz}.$$

Now differentiating with respect to t only so far as it occurs explicitly,

$$\dfrac{d}{dt}[a, b] = \dfrac{dx}{da}\dfrac{d}{dt}\left(\dfrac{dx'}{db}\right) + \dfrac{dx'}{db}\dfrac{d}{dt}\left(\dfrac{dx}{da}\right)$$

$$- \dfrac{dx}{db}\dfrac{d}{dt}\left(\dfrac{dx'}{da}\right) - \dfrac{dx'}{da}\dfrac{d}{dt}\left(\dfrac{dx}{db}\right)$$

$$+ \ldots\ldots\ldots\ldots$$

$$= \frac{dx}{da}\frac{d}{db}\left(\frac{dx'}{dt}\right) + \frac{dx'}{db}\frac{dx'}{da}$$

$$-\frac{dx}{db}\frac{d}{da}\left(\frac{dx'}{dt}\right) - \frac{dx'}{da}\frac{dx'}{db}$$

$$+ \dots\dots\dots\dots$$

$$= \frac{dx}{da}\frac{d}{db}\left(\frac{dV}{dx}\right) - \frac{dx}{db}\frac{d}{da}\left(\frac{dV}{dx}\right)$$

$$+ \frac{dy}{da}\frac{d}{db}\left(\frac{dV}{dy}\right) - \frac{dy}{db}\frac{d}{da}\left(\frac{dV}{dy}\right)$$

$$+ \frac{dz}{da}\frac{d}{db}\left(\frac{dV}{dz}\right) - \frac{dz}{db}\frac{d}{da}\left(\frac{dV}{dz}\right)$$

$$= \frac{dx}{da}\frac{d}{dx}\left(\frac{dV}{db}\right) + \frac{dy}{da}\frac{d}{dy}\left(\frac{dV}{db}\right) + \frac{dz}{da}\frac{d}{dz}\left(\frac{dV}{db}\right)$$

$$- \frac{dx}{db}\frac{d}{dx}\left(\frac{dV}{da}\right) - \frac{dy}{db}\frac{d}{dy}\left(\frac{dV}{da}\right) - \frac{dz}{db}\frac{d}{dz}\left(\frac{dV}{da}\right)$$

$$= \frac{d^2V}{da\,db} - \frac{d^2V}{db\,da} = 0.$$

Hence $[a, b]$ does not contain the time explicitly. The same is of course true of $[a, c]$, $[b, c]$, &c. It follows, then, that whatever system of elements be adopted, we can always express their differential coefficients in terms of the partial differential coefficients of R with respect to them, multiplied by coefficients which do not involve the time explicitly.

6. From the formula of Art. 3 of this Appendix, which is due to Lagrange, those of Chapter II. may be deduced: for this we refer to Pontécoulant's *Système du Monde*, Tom. I. p. 542.

APPENDIX.

ON THE GEOMETRICAL INTERPRETATION OF THE FORMULÆ FOR THE SECULAR VARIATIONS OF THE NODE AND INCLINATION.

7. In Art. 84 we have shewn that in consequence of the secular variations of the node and inclination, the normal to the plane of the orbit moves uniformly, so as to generate in space a fixed right circular cone. This result may also be obtained somewhat differently: we will here indicate the process*.

The equations to be interpreted are
$$p = \tan i \sin \Omega = N_1 \sin (h_1 t + \delta_1) + N_2 \sin \delta_2,$$
$$q = \tan i \cos \Omega = N_1 \cos (h_1 t + \delta_1) + N_2 \cos \delta_2.$$

Employing the figure and construction of Art. 84, it will be found that
$$\cot QAO = \tan i \sin \Omega, \quad \cot QLO = \tan i \cos \Omega.$$

These are the geometric representations of p and q. Now it may be shewn by Spherical Trigonometry that
$$\cot QAO = \frac{\tan I \sin \omega + \tan \rho (\sin \omega \cos \theta + \sec I \cos \omega \sin \theta)}{1 - \tan I \tan \rho \cos \theta},$$
$$\cot QLO = \frac{\tan I \cos \omega + \tan \rho (\cos \omega \cos \theta - \sec I \sin \omega \sin \theta)}{1 - \tan I \tan \rho \cos \theta}.$$

These expressions are rigorous; if we neglect the product $\tan^2 I \tan \rho$, and higher products of $\tan I$ and $\tan \rho$, they become
$$\cot QAO = \tan I \sin \omega + \tan \rho \sin (\theta + \omega),$$
$$\cot QLO = \tan I \cos \omega + \tan \rho \cos (\theta + \omega).$$

* This method as well as that of Art. 84, is due to Mr Freeman, of St John's College.

From these equations the interpretation follows as in Art. 84. The advantage of this method is that it affords a geometric representation of p and q: on the other hand, in the method of Art. 84, the trigonometrical reduction is simpler.

ON THE METHODS OF CALCULATING THE MASSES OF THE PLANETS.

8. There are in general two methods of determining the masses of the planets; either by observations on a satellite, when the planet is accompanied by a satellite; or by comparing the inequalities produced in their motion by their mutual action, as deduced from observation, with the same inequalities calculated from theory. The secular variations are best adapted to give the most exact results; but these are not yet known with sufficient accuracy to allow of this use. We are therefore obliged to recur to the periodic variations, and, by combining a vast number of observations, gather from them the most probable results*.

9. When the planet is accompanied by a satellite the formula for calculating its mass may be obtained as follows:

Let M, m, m' be the masses of the Sun, the planet, and the satellite: P, P' the periodic times of the planet about the Sun, and the satellite about the planet; a, a' the mean distances of the planet from the Sun, and the satellite from the planet. Then we have

$$P = \frac{2\pi a^{\frac{3}{2}}}{\sqrt{(M+m)}}, \quad P' = \frac{2\pi a'^{\frac{3}{2}}}{\sqrt{(m+m')}};$$

therefore
$$\frac{m+m'}{M+m} = \frac{P^2 a'^3}{P'^2 a^3},$$

---

* Pontécoulant's *Système du Monde*, Tome III. p. 340.

APPENDIX. 147

or approximately,

$$\frac{m+m'}{M} = \frac{P^2 a'^3}{P'^2 a^3}.$$

This equation gives the mass of the planet when that of its satellite is known. If the latter be neglected, the formula becomes

$$\frac{m}{M} = \frac{P^2 a'^3}{P'^2 a^3}.$$

10. In the case of the Earth, this method is not sufficiently exact, but the following may be employed. The attraction of the Earth on a body at its surface, in the parallel of which the square of the sine of the latitude is $\frac{1}{3}$, is very nearly the same as if the Earth were condensed into its centre. (See Pratt's *Figure of the Earth*, Art. 89.) Let then $\sin^2 l = \frac{1}{3}$, g = the Earth's attraction on a body at its surface in latitude l, b the mean radius of the Earth, E the mass of the Earth, M the mass of the Sun, P the length of the year, and a the mean radius of the Earth's orbit. Then

$$g = \frac{E}{b^2}, \quad P = \frac{2\pi a^{\frac{3}{2}}}{\sqrt{M}};$$

therefore

$$\frac{E}{M} = \frac{gb^2 P^2}{4\pi^2 a^3} = \frac{gP^2}{4\pi^2 b}\left(\frac{b}{a}\right)^3,$$

where $\frac{b}{a}$ = sine of Sun's parallax = $\sin 8''.57$.

11. For the methods of calculating the mass of the Moon we refer to Pontécoulant's *Système du Monde*, Tome IV. p. 651.

Tables of the numerical values of the masses of the planets, and of the elements of their orbits, will be found in Herschel's *Outlines of Astronomy*, pp. 693 et seq.

THE END.

MACMILLAN AND CO.'S
Cambridge Class Books
FOR COLLEGES AND SCHOOLS.

The Series of CAMBRIDGE CLASS-BOOKS FOR THE USE OF COLLEGES AND SCHOOLS, *which have been issued at intervals during the last ten years by* MACMILLAN AND CO., *is intended to embrace all branches of Mathematics, from the most elementary to the most advanced, and to keep pace with the latest discoveries in Mathematical science.*

Of those hitherto published the sale of many thousands is a sufficient indication of the manner in which they have been appreciated by the public.

A SERIES of a more Elementary character is in preparation, a list of which will be found at the back of this Catalogue.

WORKS BY THE REV. BARNARD SMITH, M.A.
Fellow of St. Peter's College, Cambridge.

1. Arithmetic and Algebra in their Principles and Application.
With numerous Examples, systematically arranged.

Eighth Edition. 696 pp. (1861). Crown 8vo. strongly bound in cloth. 10s. 6d.

The first edition of this work was published in 1854. It was primarily intended for the use of students at the Universities, and for Schools which prepare for the Universities. It has however been found to meet the requirements of a much larger class, and is now in use in a large number of *Schools* and *Colleges* both *at home* and in the *Colonies*. It has also been found of great service for students preparing for the MIDDLE-CLASS and CIVIL AND MILITARY SERVICE EXAMINATIONS, from the care that has been taken to elucidate the *principles* of all the Rules. Testimony of its excellence has been borne by some of the highest practical and theoretical authorities; of which the following is a specimen. The late DEAN PEACOCK may be taken as a specimen.

"Mr. Smith's Work is a most useful publication. The Rules are stated with great clearness. The Examples are well selected and worked out with just sufficient detail without being encumbered by too minute explanations; and there prevails throughout it that just proportion of theory and practice, which is the crowning excellence of an elementary work."

2. Arithmetic for the Use of Schools.
New Edition (1862) 348 pp. Crown 8vo. strongly bound in cloth, 4s. 6d. Answers to all the Questions.

3. Key to the above, containing Solutions to all the Questions in the latest Edition. Crown 8vo. cloth. 392 pp. Second Edition. 8s. 6d.

To meet a widely expressed wish, the ARITHMETIC was published separately from the larger work in 1854, with so much alteration as was necessary to make it quite independent of the ALGEBRA. It has now a large and increasing sale in all classes of Schools at home and in the Colonies. A very copious collection of Examples, under each rule, has been embodied in the work in a systematic order, and a Collection of Miscellaneous Papers in all branches of Arithmetic have been appended to the book.

4. Exercises in Arithmetic.
104 pp. Crown 8vo. (1860) 2s. Or with Answers, 2s. 6d. Also sold separately in 2 Parts price 1s. each.

Answers, 6d.

The EXERCISES have been published in order to give the pupil examples of every rule in Arithmetic, and a great number have been carefully compiled from the latest University and School Examination Papers.

2000.
Oct. 2: 62.

WORKS BY ISAAC TODHUNTER, M.A., F.R.S.

Fellow and Principal Mathematical Lecturer of St. John's College, Cambridge.

1. Algebra.

For the Use of Colleges and Schools.

Third Edition. 542 pp. (1862).
Strongly bound in cloth. 7s. 6d.

This work contains all the propositions which are usually included in elementary treatises on Algebra, and a large number of *Examples for Exercise*. The author has sought to render the work easily intelligible to students without impairing the accuracy of the demonstrations, or contracting the limits of the subject. The Examples have been selected with a view to illustrate every part of the subject, and as the number of them is about *sixteen hundred and fifty*, it is hoped they will supply ample exercise for the student. Each set of Examples has been carefully arranged, commencing with very simple exercises, and proceeding gradually to those which are less obvious.

2. Plane Trigonometry.

For Schools and Colleges.

Second Edition. 279 pp. (1860), Crn. 8vo.
Strongly bound in cloth. 5s.

The design of this work has been to render the subject intelligible to beginners, and at the same time to afford the student the opportunity of obtaining all the information which he will require on this branch of Mathematics. Each chapter is followed by a set of Examples; those which are entitled *Miscellaneous Examples*, together with a few in some of the other sets, may be advantageously reserved by the student for exercise after he has made some progress in the subject. As the Text and Examples of the present work have been tested by considerable experience in teaching, the hope is entertained that they will be suitable for imparting a sound and comprehensive knowledge of Plane Trigonometry, together with readiness in the application of this knowledge to the solution of problems. In the Second Edition the hints for the solution of the Examples have been considerably increased.

3. Spherical Trigonometry.

For the Use of Colleges and Schools.

112 pp. Crown 8vo. (1859).
Strongly bound in cloth. 4s. 6d.

This work is constructed on the same plan as the *Treatise on Plane Trigonometry*, to which it is intended as a sequel. Considerable labour has been expended on the text in order to render it comprehensive and accurate, and the Examples, which have been chiefly selected from University and College Papers, have all been carefully verified.

4. The Integral Calculus and its Applications.

With numerous Examples.

Second Edition. 342 pp. (1862).
Crown 8vo. cloth. 10s. 6d.

In writing the present *Treatise on the Integral Calculus*, the object has been to produce a work at once elementary and complete—adapted for the use of beginners, and sufficient for the wants of advanced students. In the selection of the propositions, and in the mode of establishing them, the author has endeavoured to exhibit fully and clearly the principles of the subject, and to illustrate all their most important results. In order that the student may find in the volume all that he requires, a large collection of Examples for exercise has been appended to the different chapters.

5. The Differential Calculus

With numerous Examples.

Third Edition, 398 pp. (1860).
Crown 8vo. cloth, 10s. 6d.

This work is intended to exhibit a comprehensive view of the Differential Calculus on the method of Limits. In the more elementary portions, explanations have been given in considerable detail, with the hope that a reader who is without the assistance of a tutor may be enabled to acquire a competent acquaintance with the subject. More than one investigation of a theorem has been frequently given, because it is believed that the student derives advantage from viewing the same proposition under different aspects, and that in order to succeed in the examinations which he may have to undergo, he should be prepared for a considerable variety in the order of arranging the several branches of the subject, and for a corresponding variety in the mode of demonstration.

6. Analytical Statics.

With numerous Examples.

Second Edition. 330 pp. (1858).
Crown 8vo. cloth. 10s. 6d.

In this work will be found all the propositions which usually appear in treatises on Theoretical Statics. To the different chapters Examples are appended, which have been selected principally from the University and College Examination Papers; these will furnish ample exercise in the application of the principles of the subject.

WORKS BY ISAAC TODHUNTER, M.A., F.R.S.
(Continued).

7. An Elementary Treatise on the Theory of Equations.
With a Collection of Examples.
Crown 8vo. cloth. 7s. 6d.

This treatise contains all the propositions which are usually included in elementary treatises on the Theory of Equations, together with a collection of Examples for exercise. It may be read by those who are familiar with Algebra, since no higher knowledge is assumed, except in Arts. 175, 267, 308—314, which may be postponed by those who are not acquainted with De Moivre's Theorem in Trigonometry. This work may in fact be regarded as a sequel to that on Algebra by the same writer, and accordingly the student has occasionally been referred to the treatise an Algebra for preliminary information on some topics here discussed.

8. Plane Co-Ordinate Geometry
AS APPLIED TO THE STRAIGHT LINE AND THE CONIC SECTIONS;

With numerous Examples.

Third and Cheaper Edition.
Crown 8vo. cloth. 7s. 6d.

This *Treatise* exhibits the subject in a simple manner for the benefit of beginners, and at the same time includes in one volume all that students usually require. The Examples at the end of each chapter will, it is hoped, furnish sufficient exercise, as they have been carefully selected with the view of illustrating the most important points, and have been tested by repeated experience with pupils. In consequence of the demand for the work proving much greater than had been originally anticipated, a large number of copies of the *Third Edition* has been printed, and a considerable reduction effected in the price.

....

By J. H. PRATT, M.A.
Archdeacon of Calcutta, late Fellow of Gonville and Caius College, Cambridge.

A Treatise on Attractions, La Place's Functions, and the Figure of the Earth.
Second Edition. Crown 8vo. 126 pp. (1861). cloth. 6s. 6d.

In the present Treatise the author has endeavoured to supply the want of a work on a subject of great importance and high interest—La Place's Coefficients and Functions and the calculation of the Figure of the earth by means of his remarkable analysis. No student of the higher branches of Physical Astronomy should be ignorant of La Place's analysis and its result—"a calculus," says Airy, "the most singular in its nature and the most powerful in its application that has ever appeared."

9. Examples of Analytical Geometry of Three Dimensions.
76 pp. (1858). Crown 8vo. cloth. 4s.

A collection of examples in illustration of Analytical Geometry of Three Dimensions has long been required both by students and teachers, and the present work is published with the view of supplying the want.

10. History of the Progress of the Calculus of Variations.
During the Nineteenth Century.

It is of importance that those who wish to cultivate any subject may be able to ascertain what results have already been obtained, and thus reserve their strength for difficulties which have not yet been conquered. And those who merely desire to ascertain the present state of a subject without any purpose of original investigation will often find that the study of the past history of that subject assists them materially in obtaining a sound and extensive knowledge of the condition which it has attained. The Author has endeavoured in this work to ascertain distinctly what has been effected in the Progress of the Calculus, and to form some estimate of the manner in which it has been effected: accordingly, unless the contrary is distinctly stated, it may be assumed that any treatise or memoir relating to the Calculus of Variations which is described in this work has undergone thorough examination and study.

....

By P. G. TAIT, M.A., and
W. J. STEELE, B.A.
Late Fellows of St. Peter's College, Cambridge.

Dynamics of a Particle.
With numerous Examples.

304 pp. (1856). Crown 8vo. cloth. 10s. 6d.

In this Treatise will be found all the ordinary propositions connected with the Dynamics of Particles which can be conveniently deduced without the use of D'Alembert's Principles. Throughout the book will be found a number of illustrative Examples introduced in the text, and for the most part completely worked out; others, with occasional solutions or hints to assist the student are appended to each Chapter.

By REV. S. PARKINSON, B.D.
Fellow and Prælector of St. John's Coll. Camb.

1. Elementary Treatise on Mechanics.

With a Collection of Examples.

Second Edition. 345 pp. (1861).
Crown 8vo. cloth. 9s. 6d.

The Author has endeavoured to render the present volume suitable as a Manual for the junior classes in Universities and the higher classes in Schools. With this object there have been included in it those portions of theoretical Mechanics which can be conveniently investigated without the Differential Calculus, and with one or two short exceptions the student is not presumed to require a knowledge of any branches of Mathematics beyond the elements of Algebra, Geometry, and Trigonometry. A collection of Problems and Examples has been added, chiefly taken from the Senate-House and College Examination Papers which will, it is trusted, be found useful as an exercise for the student. In the Second Edition several additional propositions have been incorporated in the work for the purpose of rendering it more complete, and the Collection of Examples and Problems has been largely increased.

2. A Treatise on Optics.

304 pp. (1859). Crown 8vo. 10s. 6d.

The present work may be regarded as a new edition of the *Treatise on Optics*, by the Rev. W. N. Griffin, which being some time ago out of print, was very kindly and liberally placed at the disposal of the author. The author has freely used the liberty accorded to him, and has rearranged the matter with considerable alterations and additions—especially in those parts which required more copious explanation and illustration to render the work suitable for the present course of reading in the University. A collection of Examples and Problems has been appended, which are sufficiently numerous and varied in character to afford an useful exercise for the student: for the greater part of them recourse has been had to the Examination Papers set in the University and the several Colleges during the last twenty years. Subjoined to the copious Table of Contents the author has ventured to indicate an elementary course of reading not unsuitable for the requirements of the First Three Days in the Cambridge Senate House Examinations.

———o———

By J. B. PHEAR, M.A.
Fellow & late Mathematical Lecturer of Clare Coll.

Elementary Hydrostatics.

With numerous Examples and Solutions.

Second Edition. 156 pp. (1857).
Crown 8vo. cloth. 5s. 6d.

"An excellent Introductory Book. The definitions are very clear; the descriptions and explanations are sufficiently full and intelligible; the investigations are simple and scientific. The examples greatly enhance its value."—ENGLISH JOURNAL OF EDUCATION.

This Edition contains 147 Examples, and solutions to all these examples are given at the end of the book.

By G. B. AIRY, M.A.
Astronomer Royal.

1. Mathematical Tracts

On the Lunar and Planetary Theories, Figure of the Earth, the Undulatory Theory of Optics, &c.

Fourth Edition. 400 pp. (1858). 8vo. 15s.

2. Theory of Errors of Observations and the Combination of Observations.

103 pp. (1861). Crown 8vo. 6s. 6d.

In order to spare astronomers and observers in natural philosophy the confusion and loss of time which are produced by referring to the ordinary treatises embracing both branches of Probabilities, the author has thought it desirable to draw up this work, relating only to Errors of Observation, and to the rules derivable from the consideration of these Errors, for the Combination of the Results of Observations. The Author has thus also the advantage of entering somewhat more fully into several points of interest to the observer, than can possibly be done in a General Theory of Probabilities.

———o———

By GEORGE BOOLE, D.C.L.
Professor of Mathematics in the Queen's University, Ireland.

1. Differential Equations.

468 pp. (1859). Crown 8vo. cloth. 14s.

The Author has endeavoured in this treatise to convey as complete an account of the present state of knowledge on the subject of Differential Equations as was consistent with the idea of a work intended, primarily, for elementary instruction. The object has been first of all to meet the wants of those who had no previous acquaintance with the subject, and also not quite to disappoint others who might seek for more advanced information. The earlier sections of each chapter contain that kind of matter which has usually been thought suitable for the beginner, while the latter ones are devoted either to an account of recent discovery, or to the discussion of such deeper questions of principle as are likely to present themselves to the reflective student in connection with the methods and processes of his previous course.

2. The Calculus of Finite Differences.

248 pp. (1860). Crown 8vo. cloth. 10s. 6d.

In this work particular attention has been paid to the connexion of the methods with those of the Differential Calculus—a connexion which in some instances involves far more than a merely formal analogy. The work is in some measure designed as a sequel to the Author's *Treatise on Differential Equations*, and it has been composed on the same plan.

By *EDWARD JOHN ROUTH, M.A.*
Fellow and Assistant Tutor of St. Peter's College, Cambridge.

Dynamics of a System of Rigid Bodies.
With numerous Examples.

336 pp. (1860). Crown 8vo. cloth. 10s. 6d.

CONTENTS: Chap. I. Of Moments of Inertia.—II. D'Alembert's Principle.—III. Motion about a Fixed Axis.—IV. Motion in Two Dimensions.—V. Motion of a Rigid Body in Three Dimensions.—VI. Motion of a Flexible String.—VII. Motion of a System of Rigid Bodies.—VIII. Of Impulsive Forces.—IX. Miscellaneous Examples.

The numerous Examples which will be found at the end of each chapter have been chiefly selected from the Examination Papers set in the University and Colleges of Cambridge during the last few years.

―――o―――

By *W. H. DREW, M.A.*
Second Master of Blackheath School.

Geometrical Treatise on Conic Sections.
With a copious Collection of Examples.

Second Edition. Crown 8vo. cloth. 4s. 6d.

In this work the subject of Conic Sections has been placed before the student in such a form that, it is hoped, after mastering the elements of Euclid, he may find it an easy and interesting continuation of his geometrical studies. With a view also of rendering the work a complete Manual of what is required at the Universities, there have been either embodied into the text, or inserted among the examples, every book-work question, problem, and rider, which has been proposed in the Cambridge examinations up to the present time.

―――o―――

By *G. H. PUCKLE, M.A.*
Principal of Windermere College.

Conic Sections and Algebraic Geometry.
With numerous Easy Examples Progressively arranged.

Second Edition. 264 pp. (1856). Crown 8vo. 7s. 6d.

This book has been written with special reference to those difficulties and misapprehensions which commonly beset the student when he commences. With this object in view, the earlier part of the subject has been dwelt on at length, and geometrical and numerical illustrations of the analysis have been introduced. The Examples appended to each section are mostly of an elementary description. The work will, it is hoped, be found to contain all that is required by the upper classes of schools and by the generality of students at the Universities.

By *N. M. FERRERS, M.A.*
Fellow and Mathematical Lecturer of Gonville and Caius College, Cambridge.

An Elementary Treatise on Trilinear Co-Ordinates.
The Method of Reciprocal Polars, and the Theory of Projectiles.

154 pp. (1861). Crown 8vo. cloth. 6s. 6d.

The object of the Author in writing on this subject has mainly been to place it on a basis altogether independent of the ordinary Cartesian System, instead of regarding it as only a special form of abridged Notation. A short chapter on Determinants has been introduced.

―――o―――

By *R. D. BEASLEY, M.A.*
Head Master of Grantham School.

Plane Trigonometry.
AN ELEMENTARY TREATISE.
With a numerous Collection of Examples.

106 pp. (1858), strongly bound in cloth. 3s. 6d.

This Treatise is specially intended for use in Schools. The choice of matter has been chiefly guided by the requirements of the three days' Examination at Cambridge, with the exception of proportional parts in logarithms, which have been omitted. About four hundred examples have been added, mainly collected from the Examination Papers of the last ten years, and great pains have been taken to exclude from the body of the work any which might dishearten a beginner by their difficulty.

―――o―――

By *J. C. SNOWBALL, M.A.*
Late Fellow of St. John's College, Cambridge.

Plane and Spherical Trigonometry.
With the Construction and Use of Tables of Logarithms.

Ninth Edition. 240 pp. (1857). Crown 8vo. 7s. 6d.

In preparing a new edition, the proofs of some of the more important propositions have been rendered more strict and general; and a considerable *addition, of more than two hundred examples*, taken principally from the questions in the Examinations of Colleges and the University, has been made to the collection of Examples and Problems for practice.

By B. DRAKE, M.A.
Late Fellow of King's College, Cambridge.

1. Demosthenes on the Crown.
With English Notes.
Second Edition. To which is prefixed ÆSCHINES AGAINST CTESIPHON. With English Notes.
287 pp. (1860). Fcap. 8vo. cloth. 5s.

The first edition of the late Mr. Drake's edition of Demosthenes de Corona having met with considerable acceptance in various Schools, and a new edition being called for, the Oration of Æschines against Ctesiphon, in accordance with the wishes of many teachers, has been appended with useful notes by a competent scholar.

2. Æschyli Eumenides.
With English Verse Translation, Copious Introduction, and Notes.

"Mr. Drake's ability as a critical Scholar is known and admitted. In the edition of the Eumenides before us we meet with him also in the capacity of a Poet and Historical Essayist. The translation is flowing and melodious, elegant and scholarlike. The Greek Text is well printed: the notes are clear and useful."—GUARDIAN.

———o———

By J. BROOK SMITH, M.A.
St. John's College, Cambridge.

Arithmetic in Theory and Practice.
For Advanced Pupils.
PART I. Crown 8vo. cloth. 3s. 6d.

This work forms the first part of a Treatise on Arithmetic, in which the Author has endeavoured, from very simple principles, to explain in a full and satisfactory manner all the important processes in that subject.
The proofs have in all cases been given in a form entirely arithmetical: for the author does not think that recourse ought to be had to Algebra until the arithmetical proof has become hopelessly long and perplexing.
At the end of every chapter several examples have been worked out at length, in which the best practical methods of operation have been carefully pointed out.

———o———

By the Rev. G. F. CHILDE, M.A.
Mathematical Professor in the South African Coll.

Singular Properties of the Ellipsoid
And Associated Surfaces of the Nth *Degree.*
152 pp. (1861). 8vo. boards. 10s. 6d.

As the title of this volume indicates, its object is to develope peculiarities in the Ellipsoid; and further, to establish analogous properties in unlimited congeneric series of which this remarkable surface is a constituent.

By C. W. UNDERWOOD, M.A.
Vice-Principal of the Collegiate Institution Liverpool.

A Short Manual of Arithmetic.
Fcp. 8vo. 96 pp. (1860). limp cloth. 2s. 6d.

The object aimed at by the Compiler of this *Manual* is to bring before *junior* students so much of the Theory of Arithmetic as may be fairly expected of them, and to present it in such a form that the study of the Science may become to some extent a *mental training*. It is rather a *Grammar* of Arithmetic than a treatise on that subject, and should for the most part be committed to memory. It will be found well adapted for *vivâ voce* examination, and enable candidates to prepare themselves for the Local University Examinations. The Definitions are briefly and carefully worded. Each rule is stated so as to include the proof of it where this was possible.

———o———

Senate-House Mathematical Problems.
With Solutions.

1848-51. By FERRERS and JACKSON. 8vo. 15s. 6d.
1848-51. (RIDERS). By JAMESON. 8vo. 7s. 6d.
1854. By WALTON and MACKENZIE. 8vo. 10s. 6d.
1857. By CAMPION and WALTON. 8vo. 8s. 6d.
1860. By ROUTH and WATSON. Crown 8vo. 7s. 6d.

The above books contain Problems and Examples which have been set in the Cambridge Senate-house Examinations at various periods during the last twelve years, together with Solutions of the same. The Solutions are in all cases given by the Examiners themselves or under their sanction.

———o———

By H. A. MORGAN, M.A.
Fellow of Jesus College, Cambridge.

A Collection of Mathematical Problems and Examples, with Answers.
190 pp. (1858). Crown 8vo. 6s. 6d.

This book contains a number of problems, chiefly elementary, in the Mathematical subjects usually read at Cambridge. They have been selected from the papers set during late years at Jesus College. Very few of them are to be met with in other collections, and by far the larger number are due to some of the most distinguished Mathematicians in the University.

By JOHN E. B. MAYOR, *M.A.*
Fellow and Classical Lecturer of St. John's College, Cambridge.

1. Juvenal.

With English Notes.

464 pp. (1854). Crown 8vo. cloth. 10*s.* 6*d.*

"A School edition of Juvenal, which, for really ripe scholarship, extensive acquaintance with Latin literature, and familiar knowledge with Continental criticism, ancient and modern, is unsurpassed, we do not say among English School-books, but among English editions generally."—EDINBURGH REVIEW.

2. Cicero's Second Philippic

With English Notes.

168 pp. (1861). Fcp. 8vo. cloth. 5*s.*

The Text is that of Halm's 2nd edition, (Leipsig, Weidmann, 1858), with some corrections from Madvig's 4th Edition (Copenhagen, 1858). Halm's Introduction has been closely translated, with some additions. His notes have been curtailed, omitted, or enlarged, at discretion; passages to which he gives a bare reference, are for the most part printed at length; for the Greek extracts an English version has been substituted. A large body of notes, chiefly grammatical and historical, has been added from various sources. A list of books useful to the student of Cicero, a copious Argument, and an Index to the introduction and notes, complete the book.

―――o―――

By C. MERIVALE, *B.D.*
Author of "A History of Rome," &c.

Sallust.

With English Notes.

Second Edition. 172 pp. (1858). Fcap. 8vo. 4*s.* 6*d.*

"This School edition of Sallust is precisely what the School edition of a Latin author ought to be. No useless words are spent in it, and no words that could be of use are spared. The text has been carefully collated with the best editions. It is printed in a large bold type, which manifests a just regard for the young eyes that are to work upon it: under the text there flows through every page a full current of extremely well-selected annotations."—THE EXAMINER.

The "CATILINA" and "JUGURTHA" *may be had separately, price* 2*s.* 6*d. each, bound in cloth.*

The Cambridge Year Book

AND UNIVERSITY ALMANACK

For 1862.

Crown 8vo. 228 pp. price 2*s.* 6*d.*

The specific features of this annual publication will be obvious at a glance, and its value to teachers engaged in preparing students for, and to parents who are sending their sons to, the University, and to the public generally will be clear.

1. The whole mode of proceeding in entering a student at the University and at any particular College is stated.

2. The course of the studies as regulated by the University examinations, the manner of these examinations, and the specific subjects and times for the year 1862 are given.

3. A complete account of all Scholarships and Exhibitions at the several Colleges, their value, and the means by which they are gained.

4. A brief summary of all Graces of the Senate, Degrees conferred during the year 1861, and University news generally are given.

5. The Regulations for the LOCAL EXAMINATION of those who are not members of the University, to be held this year, with the names of the books on which the Examination will be based, and the date on which the Examination will be held.

―――o―――

Cambridge University Examination Papers, 1860-1.

A Collection of all the Papers Set at the Examinations for the Degrees, the various Triposes, and the Theological Certificates in the University, with List of Candidates Examined and of those Approved, and an Index to the Subjects.

Previous Examination, 1860—Previous Examination, 1861—B.A. Degree Examination, 1860—B.A. Degree Examination, Jan. and May, 1861—Bachelor of Laws Examination, 1860—Bachelor of Laws Examination, 1861—Bachelor of Medicine Examinations, 1861—Classical Tripos, 1861—Moral Sciences Tripos, 1861—Natural Sciences Tripos, 1861—Smith's Prizes, 1861—Chancellor's Medals for Legal and Classical Studies, 1861.

By J. HERBERT LATHAM, M.A.
Civil Engineer.

The Construction of Wrought-Iron Bridges.

Embracing the Practical Application of the Principles of Mechanics to Wrought-Iron Girder-Work.

"The great merit of this book is that it deals with practice more than theory. All the calculations in the book connected with the strength of girders are based upon their actual application which abounds in practical investigations into girder-work in all its bearings, and will be welcomed as *one of the most valuable contributions yet made to this important branch of engineering.*"
—ATHENÆUM.

———o———

By J. WRIGHT, M.A.
Head Master of Sutton Coldfield School.

1. Help to Latin Grammar.

With Easy Exercises, both English and Latin, Questions and Vocabulary.

175 pp. (1855). Crown 8vo. cloth. 4s. 6d.

"This book aims at helping the learner to overstep the threshold difficulties of the Latin Grammar; and never was there a better aid offered alike to teacher and scholar in that arduous pass. The style is at once familiar and strikingly simple and lucid; and the explanations precisely hit the difficulties, and thoroughly explain them. It will also much facilitate the acquirement of English Grammar."—ENGLISH JOURNAL OF EDUCATION.

2. The Seven Kings of Rome.

A First Latin Reading Book, abridged from Livy, by the omission of difficult passages, with Notes and Index.

Second Edition, 138 pp. (1857). Fcap. 8vo. cloth. 3s.

This work is intended to supply the pupil with an easy Construing-book, which may, at the same time, be made the vehicle for instructing him in the rules of grammar and principles of composition. These branches of the study of Latin seem to the author to have hitherto been kept too much apart. Boys have construed their Delectus, or Eutropius, or Nepos, and have gone elsewhere for their grammatical exercises. Nor can this be wondered at. An educated man must feel positively ashamed of taking his pupils away from our good English authors, and setting before him instead a Delectus or Eutropius. He therefore skims over them as lightly, and escapes from them as quickly as possible, and has recourse for his composition lesson to one of the many Exercise-books which swarm from our educational press. To remedy these evils this book has been published. Here Livy tells his own pleasant stories in his own pleasant words. What is omitted, is that which no one can wish a beginner to learn, and which may be better learnt elsewhere. Let Livy be the master to teach a boy Latin, not some English collector of sentences, and he will not be found a dull one.

3. Hellenica.

A FIRST GREEK READING BOOK.
From Diodorus and Thucydides. With Vocabulary.

Second Edition. 150 pp. (1851). Fcap. 8vo. cloth. 3s. 6d.

In the last twenty chapters of this volume, Thucydides sketches the rise and progress of the Athenian Empire in so clear a style and in such simple language, that the author doubts whether any easier or more instructive passages can be selected for the use of the pupil who is commencing Greek.

4. Vocabulary and Exercises on "The Seven Kings of Rome."

94 pp. (1857). Crown 8vo. cloth. 2s. 6d.

The Vocabulary is published apart from the Text in order to suit the views of those who prefer their pupils to consult a general dictionary, but it may also be had bound together with the "SEVEN KINGS OF ROME," if preferred. As the aim of the Text is to teach the elements of grammar, so the Exercises are intended to test the Pupil's knowledge of grammar. Indeed there is hardly an ordinary Latin construction which is not illustrated in the text, explained in the notes, and proved in the Exercises.

———o———

Elementary Statics.

By the Rev. GEORGE RAWLINSON,
Professor of Applied Sciences, Elphinstone Coll., Bombay.

Edited by the Rev. E. STURGES, M.A.
Rector of Kencott, Oxfordshire.

(150 pp.) 1860. Crown 8vo. cloth. 4s. 6d.

This work is published under the authority of H. M. Secretary of State for India for use in the Government Schools and Colleges in India.

———o———

By P. FROST, Jun., M.A.
Late Fellow of St. John's College, Cambridge.

Thucydides. Book VI.

With English Notes, Map and Index.

8vo. cloth. 7s. 6d.

It has been attempted in this work to facilitate the attainment of accuracy in translation. With this end in view the Text has been treated grammatically.

By EDWARD THRING, M.A.
Head Master of Uppingham Grammar School.

1. Elements of Grammar Taught in English.

With Questions.
Third Edition. 136 pp. (1860). Demy 18mo. 2s.

2. The Child's English Grammar.

New Editton. 86 pp. (1859). Demy 18mo. 1s.

The Author's effort in these two books has been to point out the broad, beaten, every-day path, carefully avoiding digressions into the byeways and eccentricities of language. This Work took its rise from questionings in National Schools, and the whole of the first part is merely the writing out in order the answers to questions which have been used already with success. The study of Grammar in English has been much neglected, nay by some put on one side as an impossibility. There was perhaps much ground for this opinion, in the medley of arbitrary rules thrown before the student, which applied indeed to a certain number of instances, but would not work at all in many others, as must always be the case when principles are not put forward in a language full of ambiguities. The present work does not, therefore, pretend to be a compendium of idioms, or a philological treatise, but a Grammar. Or in other words, its intention is to teach the learner how to speak and write correctly, and to understand and explain the speech and writings of others. Its success, not only in National Schools, from practical work in which it took its rise, but also in classical schools, is full of encouragement,

3. A First Latin Construing Book.

104 pp. (1855). Fcap. 8vo. 2s. 6d.

This Construing Book is drawn up on the same sort of graduated scale as the Author's *English Grammar*. Passages out of the best Latin Poets are gradually built up into their perfect shape. The few words altered, or inserted as the passages go on, are printed in Italics. It is hoped by this plan that the learner, whilst acquiring the rudiments of language, may store his mind with good poetry and a good vocabulary.

4. School Songs.

A COLLECTION OF SONGS FOR SCHOOLS.

WITH THE MUSIC ARRANGED FOR FOUR VOICES.

Edited by Rev. E. THRING & H. RICCIUS. Music Size. 7s. 6d.

Religious Class Books

By C. J. VAUGHAN, D.D.
Head Master of Harrow School.

St. Paul's Epistle to the Romans.

The Greek Text with English Notes.

Second Edition. Crown 8vo. cl. (1861). 5s.

By dedicating this work to his elder Pupils at Harrow, the Author hopes that he sufficiently indicates what is and what is not to be looked for in it. He desires to record his impression, derived from the experience of many years, that the Epistles of the New Testament, no less than the Gospels, are capable of furnishing useful and solid instruction to the highest classes of our Public Schools. If they are taught accurately, not controversially; positively, not negatively; authoratively, yet not dogmatically; taught with close and constant reference to their literal meaning, to the connexion of their parts, to the sequence of their argument, as well as to their moral and spiritual instruction; they will interest, they will inform, they will elevate; they will inspire a reverence for Scripture never to be discarded, they will awaken a desire to drink more deeply of the Word of God, certain hereafter to be gratified and fulfilled.

———o———

Notes for Lectures on Confirmation: With Suitable Prayers. By C. J. VAUGHAN, D.D. Fourth Edition. 70 pp. (1862). Fcp. 8vo. 1s. 6d.

The Church Catechism Illustrated and Explained. By ARTHUR RAMSAY, M.A. 18mo. cloth.

Hand-Book to Butler's Analogy. By C. A. SWAINSON, M.A. 55 pp. (1856). Crown 8vo. 1s. 6d.

History of the Christian Church during the First Three Centuries, and the Reformation in England. By W. SIMPSON, M.A. 307 pp. (1857). Fcp. 8vo. cloth. 5s.

Analysis of Paley's Evidences of Christianity. By CHARLES H. CROSSE, M.A. 115 pp. (1855). 18mo. 3s. 6d.

Manuals for Theological Students.

UNIFORMLY PRINTED AND BOUND.

This Series of Theological Manuals has been published with the aim of supplying Books concise, comprehensive, and accurate, convenient for the Student, and yet interesting to the general reader.

1. **History of the Christian Church during the Middle Ages.** By ARCHDEACON HARDWICK. Second Edition. 482 pp. (1861). With Maps. Crown 8vo. cloth. 10s. 6d.

   This Volume claims to be regarded as an integral and independent treatise on the Mediæval Church. The History commences with the time of Gregory the Great, because it is admitted on all hands that his pontificate became a turning-point, not only in the fortunes of the Western tribes and nations, but of Christendom at large. A kindred reason has suggested the propriety of pausing at the year 1520,—the year when Luther, having been extruded from those Churches that adhered to the Communion of the Pope, established a provisional form of government and opened a fresh era in the history of Europe.

2. **History of the Christian Church during the Reformation.** By ARCHDEACON HARDWICK. 459 pp. (1856). Crown 8vo. cloth. 10s. 6d.

   This Work forms a Sequel to the Author's Book on The Middle Ages. The Author's wish has been to give the reader a trustworthy version of those stirring incidents which mark the Reformation period.

3. **History of the Book of Common Prayer.** With a Rationale of its Offices. By FRANCIS PROCTER, M.A. Fifth Edition. 464 pp. (1860). Crown 8vo. cloth. 10s. 6d.

   The Subject of this Book has been already treated by numerous writers of distinction. When the present series of Manuals was projected, it did not appear that any one of the existing volumes taken singly was available for the desired object, in the course of the last twenty years the whole question of liturgical knowledge has been reopened with great learning and accurate research, and it is mainly with the view of epitomizing their extensive publications, and correcting by their help the errors and misconceptions which had obtained currency, that the present volume has been put together.

4. **History of the Canon of the New Testament during the First Four Centuries.** By BROOKE FOSS WESTCOTT, M.A. 594 pp. (1855). Crown 8vo. cloth. 12s. 6d.

   The Author has endeavoured to connect the history of the New Testament Canon with the growth and consolidation of the Catholic Church, and to point out the relation existing between the amount of evidence for the authenticity of its component parts and the whole mass of Christian literature. Such a method of inquiry will convey both the truest notion of the connexion of the written Word with the living Body of Christ, and the surest conviction of its divine authority.

5. **Introduction to the Study of the GOSPELS.** By BROOKE FOSS WESTCOTT, M.A. 458 pp. (1860). Crown 8vo. cloth. 10s. 6d.

   The title of this book will explain the chief aim which the Author had in view. It is intended to be an Introduction to the *Study* of the Gospels. The Author has therefore confined himself in many cases to the mere indication of lines of thought and inquiry from the conviction that truth is felt to be more precious in proportion as it is opened to us by our own work. In a subject which involves so vast a literature much must have been overlooked; but the Author has made it a point at least to study the researches of the great writers, and consciously to neglect none.

BOOKS SUITABLE FOR SCHOOL PRIZES.
KEPT IN VARIOUS BINDINGS BY THE PUBLISHERS.

THE GOLDEN TREASURY OF THE BEST SONGS AND LYRICAL POEMS IN THE ENGLISH LANGUAGE. Cloth, 4s. 6d., morocco plain, 7s. 6d., extra, 10s. 6d.

THE CHILDREN'S GARLAND FROM THE BEST POETS. Cloth, 4s. 6d. morocco plain, 7s. 6d., extra, 10s. 6d.

BUNYAN'S PILGRIM'S PROGRESS. Cloth, 4s. 6d., morocco plain, 7s. 6d., extra, 10s. 6d.

TOM BROWN'S SCHOOL DAYS. By An Old Boy. Seventh Edition. Fcap. 8vo. cloth, 5s., calf, 9s., morocco, 11s.

THE HEROES; or GREEK FAIRY TALES. By Charles Kingsley, Rector of Eversley. Second Edition, with Eight Illustrations. Imperial 16mo. cloth, gilt leaves, 5s., calf, 9s. 6d., morocco, 12s.

DAVID, KING OF ISRAEL; Readings for the Young. With Six Illustrations. By J. Wright, M.A. Imp. 16mo. cloth, 5s., calf, 9s. 6d., mor. 12s.

LITTLE ESTELLA AND OTHER FAIRY TALES. Imp. 16mo. cloth, 5s., calf, 9s. 6d., mor. 12s.

PAULI'S PICTURES OF OLD ENGLAND. Crown 8vo. cloth, 8s. 6d., calf, 13s. 6d., mor. 16s. 6d.

DAYS OF OLD: STORIES FROM OLD ENGLISH HISTORY of the Druids, Anglo-Saxons, and the Crusaders. By the Author of Ruth and Her Friends. Cloth, 5s., calf, 9s. 6d., mor. 12s.

RUTH AND HER FRIENDS. A Story for Girls. Third Edition. Cloth, 5s., calf, 9s. 6d., mor. 12s.

OUR YEAR: A Child's Book in Prose and Verse. By the Author of "John Halifax." Numerous Illustrations. Royal 16mo. cloth, gilt leaves, 5s., calf, 9s. 6d., mor. 12s.

EARLY EGYPTIAN HISTORY FOR THE YOUNG. With Description of the Tombs and Monuments. Fcp. 8vo. cloth, 5s., calf, 9s. 6d., mor. 12s.

WESTWARD HO! THE ADVENTURES OF SIR AMYAS LEIGH in the Reign of Elizabeth. Third Edition. By Charles Kingsley. Crown 8vo. cloth, 6s. calf, 11s., mor. 14s.

TWO YEARS AGO. By Charles Kingsley. Third Edition. Crown 8vo. cloth, 6s., calf, 11s., mor. 14s.

THE RECOLLECTIONS OF GEOFFRY HAMLYN. By Henry Kingsley. Second Edition. Crown 8vo. cloth, 6s., calf, 11s., mor. 14s.

GLAUCUS; or, WONDERS OF THE SHORE. By Charles Kingsley. Illustrated Edition, containing Coloured Illustrations of the objects mentioned in the Work. Imp. 16mo. cloth, gilt leaves, 7s. 6d., calf, 12s., mor. 14s. 6d.

ESSAYS, CHIEFLY ON ENGLISH POETS. By David Masson, M.A. 8vo. cloth, 12s. 6d., calf, 18s., mor. 22s. 6d.

ARCHER BUTLER'S HISTORY OF ANCIENT PHILOSOPHY. 2 vols. 8vo. cloth, 1l. 5s., calf, 1l. 16s., mor. 2l. 5s.

HISTORY OF THE CANON OF THE NEW TESTAMENT. By B. F. Westcott, M.A. Third Edition. Crown 8vo. cloth, 12s. 6d., calf, 17s. 6d., mor. 1l. 6s.

INTRODUCTION TO THE STUDY OF THE GOSPELS. By B. F. Westcott, M.A. Crown 8vo. cloth, 10s. 6d., calf, 15s. 6d., mor. 18s. 6d.

GEORGE BRIMLEY'S ESSAYS. Second Edition. Fcp. 8vo. cloth, 5s., calf, 9s., mor. 11s.

THE PLATONIC DIALOGUES FOR ENGLISH READERS. By W. Whewell, D.D. 3 vols. Fcp. 8vo. cloth. 21s. 6d., calf, 35s., mor. 42s. 6d.

MACMILLAN & CO.'S
ELEMENTARY SCHOOL CLASS BOOKS.

Euclid for Colleges and Schools.
By I. TODHUNTER, M.A., F.R.S., Fellow and Principal Mathematical Lecturer of St. John's College, Cambridge. [*Nearly Ready.*

⁎ This work, which will form a volume of Macmillan and Co.'s ELEMENTARY SCHOOL CLASS BOOKS, will be handsomely printed in 18mo. and all the volumes of the series will be published at a low price, to ensure an extensive sale in the Schools of the United Kingdom and the Colonies.

The following will form early volumes in this Series:

An Elementary History of the Prayer-Book.
By FRANCIS PROCTER, M.A., Vicar of Witton, Norfolk, late Fellow of St. Catharine's College, Cambridge. [*Nearly Ready.*

The School Class Book of Arithmetic.
By BARNARD SMITH, M.A., Fellow of St. Peter's College, Cambridge.

Elementary Latin Grammar.
By H. J. ROBY, M.A., Assistant Master in Dulwich College, formerly Fellow of St. John's College.

The Bible Word-Book.
A Glossary of old English Bible Words, with Illustrations. By J. EASTWOOD, M.A., St. John's College, Cambridge, and W. ALDIS WRIGHT, M.A., Trinity College, Cambridge.

⁎ Other Volumes will be announced in due course.

FORTHCOMING BOOKS.

An Elementary Treatise on the Planetary Theory.
By C. H. H. CHEYNE, B.A., Scholar of St. John's College. [*In the Press.*

An Elementary Treatise on Natural Philosophy.
By WILLIAM THOMSON, LL.D., F.R.S., late Fellow of St. Peter's College, Cambridge, Professor of Natural Philosophy in the University of Glasgow; and PETER GUTHRIE TAIT, M.A., late Fellow of St. Peter's College, Cambridge, Professor of Natural Philosophy in the University of Edinburgh. With numerous Illustrations. [*Preparing.*

An Elementary Treatise on Quaternions.
With numerous Examples. By P. G. TAIT, M.A., Professor of Natural Philosophy in the University of Edinburgh. [*Preparing.*

A Treatise on Geometry of Three Dimensions.
By PERCIVAL FROST, M.A., St. John's College, and JOSEPH WOLSTEN-HOLME, M.A., Christ's College, Cambridge. [*In the Press.*

⁎ The First Portion has been issued for the convenience of Cambridge Students.

First Book of Algebra. For Schools.
By J. C. W. ELLIS, M.A., and P. M. CLARK, M.A., Sidney Sussex College, Cambridge. [*Preparing.*

The New Testament in the Original Greek.
Text revised by B. F. WESTCOTT, M.A., and F. J. HORT, M.A., formerly Fellows of Trinity College.

Aristotelis de Rhetorica.

www.ingramcontent.com/pod-product-compliance
Lightning Source LLC
Chambersburg PA
CBHW020305170426
43202CB00008B/504